PAPER
CUP
DESIGN NOW!

纸杯设计

（波兰）伊维莉娜·柏臣 / 编 贺丽 / 译

辽宁科学技术出版社

PREFACE

前言

Graphic designer's aim is to find the shortest and the most effective way to present art to the mass audience, and find its understanding and acceptance. Although it may seem to be a very easy task, balancing between art and commercial works requires a lot of knowledge and empathy. These two are necessary to succeed. As a designer, you need to be not only the artist who creates something, but also the consumer. You need to fit to the particular age group, or some particular culture in nowadays society. You simply need develop in yourself two opposite aspects with the same strength, and help them to achieve agreement. It's like a battle between heart and reason. Like effort to achieve something that from the beginning has now bright future, because the winner can be only one. And then comes the hardest part – in design aspects must stay the same important, none can lose, none can win. Creating the perfect balance is the result of deducting the same amounts from both sides, resigning the same much on both sides. Final decision of this process is the ready product that can be presented to the audience. But to find its interest, design must not only be useful, but also have something original or surprising, something that will make people remember it. Expectations and esthetic needs of consumers become with every day more and more sophisticated. They start to pay attention to the details. They appreciate the nice packaging, interesting form, smart visual solutions. This growth of esthetic consciousness is the result of commonly existing and improving with every day, graphic design which surrounds us. Posters, bags, paper cups, packaging, everything is designed to meet the expectations of the group that its addresed to.

平面设计师的首要目标就是寻找到一种最简短且最有效的方式来将艺术呈献给大众群体，并且能够使其被理解和接受。尽管在艺术与商业作品之前寻求平衡点看起来是一份非常简单的任务，却也需要大量的知识与情感的投入。这两者是成功必不可少的元素。作为设计师，有时不能仅仅把自己当作是设计某种作品的艺术家，同样需要站在消费者的角度来思考。你需要去使你的设计能够适应一些特定的年龄群体，或是当今社会存在的一些特殊文化。你要做的仅仅是在这两个方面都下同样大的功夫来提升自己，并且得到认可。这就像是一场感性与理性的战争。就像是一件从头做起的事情经过努力终于看见了光明的未来一样，因为最后的胜者只有一个。接踵而来的是最为艰难的部分——在设计的各个方面都需要同样的重视，没有孰轻孰重之分。有时为了完美的保持这种平衡，往往需要两者都做出一些删减与妥协。而这个过程的最终结果就是呈现给观众的产品设计。但有趣的是，设计的价值一定不能仅仅只是实用性，同样需要具有某种原创性和惊喜在里面，一些能够让人们记住你的东西。而顾客的期望值和美学感官的需求也日益增加。他们现在会更加的关注于细节的部分，希望看到更加漂亮的包装，更有趣的形式和更加完善的视觉效果。我们生活中必不可少并且又在日复一日中不断增多的平面设计直接致使了人们美学意识的增强。海报，包袋，纸质水杯，包装袋，一切的设计都是为了迎合它们的受众群体。

This book is presenting the Paper Cup design as one of the most popular, omnipresent products that exist in our lives. It's the attribute familiar to 100% of our society and for 80% of them is an integral part of their mornings. Let's imagine... 9 am, Monday, Brussels, crowded streets, everyone is on the way to his job. Almost each one carries a paper cup. They are from different café bars, chain cafés, bakeries, but everyone has it! Fancy ladies carrying briefcases, carry stylish cups with elegant design, from the chick bakery; Young girls carry ecologic cups, from chain cafés, with light images but very interesting and comfortable handles... All the aspects like: lifestyle, food preferences, average age of the target, expectations, are extremely important in creating good design. These aspects influence our decisions about the material, type of construction, type of pattern, colours, importance of the brand... Therefore, the book that you are holding right now, gathers all the factors that are important in designing paper cups. You can find here the selection of the great works that may be an inspiration, and also solid explanation of the technical part of the design.

Inspire yourself! And enjoy!

Ewelina Bocian

《纸杯设计》这本书向读者呈现的正是目前最流行的，也是我们生活中无处不在的产品。纸杯与我们当今社会的契合度几乎是百分之百的，其中百分之八十又是每天清晨不可或缺的部分。让我们想象一下……布鲁塞尔的星期一早上9点钟，拥挤的街道，人们都奔走在上班的路上，几乎每个人手里都拿着一个纸杯。尽管这些纸杯来自于不同的咖啡厅，连锁咖啡店，面包店，但相同的是每个人手中都有一个。迷人的女士们提着公文包，手里端着面包店的设计优雅的杯子。年轻的女孩端着咖啡厅的环保纸杯，上面画着浅浅的图案，非常有趣又方便携带……所有这些方面如：生活方式，所喜爱的食物，目标年龄层，期望值，这一切都对优秀的设计至关重要。这些方面直接影响了我们对原料的选用，生产的方式的选取，图案的绘制和颜色的使用，和对于一个品牌最重要的……简约的设计。因此，这本你正在捧读的书，集合了一切的纸杯设计方面的重要元素，同样也提供了可靠的解决设计中涉及到的技术性问题的方案。

但愿你能从中得到启发！请享受愉快的阅读过程吧！

伊维莉娜·柏臣

CONTENTS
目录

Definition of Paper Cup
纸杯的概念

In this modern society where convenience, health, safety and environmental protection turn to be important, disposable cups become the mainstream. Also the disposable paper cups gradually take the place of disposable plastic cups and become the market leader. What's more, they also gradually enter into the catering industry and instead of cutlery containers, being the common drinking tools in many homes and public places. The paper cup develops from the conical / pleated paper cup, waxed paper cup and the straight double-wall cup to today's laminated plastic coated paper cup.

The original paper cup was conical and made up by means of gluing manually. It was not so tight and should be used as soon as possible. Then the folding paper cups emerged, in which the pleats were added to enhance the strength of sidewall as well as the durability of paper cups. While printing pattern on the folded surface is quite difficult, even the final effect is unsatisfying. In 1932, the first waxed cup came out, whose smooth surface can be printed on a variety of exquisite patterns, improving its marketing effectiveness. Wax could isolate drinks from the paper material and protect cups, enhancing the durability of the paper cups; on the other hand, it would increase the thickness of the sidewall, greatly improving the strength of the paper cups, thereby reducing the amount of paper consumption and decreasing the production costs. Following the waxed paper cup for cold drinks, a convenient container to hold hot drinks was expected to be created. However, hot drinks would melt the surface layer of wax and cause the gluing joint open; hence, the wax paper cups do not suit for containing hot drinks. In order to expand the application scope of the cups, in 1940, straight double-wall cup was introduced into the market. It was not only easy to take, but could contain the hot drinks. Some food companies began to paint the polyethylene onto cardboard, in order to increase the insulativity and tightness of paper packaging. Its melting point is much higher than that of wax, and the new drink cups coated with this material is ideal to contain hot drinks. Meanwhile, the polyethylene coating is smoother than that of the original wax, improving the appearance of the cups. This kind of paper cup had created a new world.

在人们追求方便、卫生、安全以及环保的现代社会，一次性杯子在人们的日常生活中逐渐体现其优越性。并且随着一次性塑料杯的逐步退出现代消费市场，一次性纸杯成为现代社会一次性杯子的主流。它也逐步进入餐饮行业代替餐具容器，是许多家庭和公共场所常见的饮水工具。纸杯是由圆锥形/折褶纸杯、涂蜡纸杯和直壁双层杯逐步发展到今天的淋膜涂塑纸杯。

最初问世的纸杯是圆锥形的，由手工制造，用胶粘合，比较容易开，必须尽快使用。随后出现了折叠纸杯，在侧壁加了折褶，以增加侧壁的强度和纸杯的耐用性。但要在这些折叠表面印上图案就较为困难，而且效果不甚理想。1932年，第一只涂蜡两片纸杯问世，其平滑的表面可以印上各种精美图案，提高促销效果。纸杯涂蜡，可以避免饮品与纸材直接接触，能保护纸杯，增强纸杯的耐用性；另一方面也增加了侧壁的厚度，使纸杯的强度大大提高。从而减少制造较结实的纸杯所必需的纸用量，降低生产成本。随着涂蜡纸杯成为冷饮品的盛装器皿，人们也希望能有一种方便的器皿来盛装热饮。但是，热饮会融化纸杯内表面的蜡层，胶口也会分离，所以，涂蜡纸杯并不适用于盛载热饮料。为扩大纸杯的应用范围，1940年，直壁双层纸杯被推向市场。这种纸杯不仅便于携带，还可用以盛装热饮料。一些食品公司开始将聚乙烯涂在纸板上，以增加纸包装阻隔性和密封性。它的熔点大大高于蜡质，采用这种材料涂布的新型饮料纸杯，能理想地用以盛载热饮料。同时，聚乙烯涂料比原先的蜡涂料平滑，改进了纸杯的外观。这为纸杯的设计开创了一片新的天地。

This ice cream paper cup is designed for Pinkberry brand. It has fully presented the characteristics of Pinkberry's products. A flower folded with a ribbon is attached on the side of paper cup; along with the balloon, the logo of Pinkberry sits at the back of cup has added the cup with a sense of cuteness and deliciousness.

这是为Pinkberry品牌设计的冰淇淋纸杯。这个纸杯充分体现了Pinkberry的产品特点，纸杯上面有一朵丝带叠成的小花，纸杯后面还附带印有Pinkberry企业标志的气球，更显得精致可爱，十分美味诱人。

These cups are designed for Gaufres & Goods rustic restaurant. The cups printed with the rustic but cosy patterns will give people a sense of warmness, strengthening the brand's characteristics of the restaurant.

这是为Gaufres & Goods乡村餐厅设计的纸杯。该纸杯印有乡村风格的温馨图案，让人有一种温暖的感觉，更加强了该餐厅品牌的特点。

These cups are designed by João Ricardo Machado for Yummy ice cream. This is specially designed for children, and vibrant colours as well as lovely expression on the cups could easily attract children's attention.

这是João Ricardo Machado为Yummy冰淇淋设计的纸杯。该产品专为儿童设计，鲜明的色彩以及纸杯上可爱的表情特别能吸引儿童的目光。

Types of Paper Cup
纸杯的种类

With the rapid development of socio-economic, the method of cup designing is always improved, which not only meets the aesthetic needs of people but lays more focus on the functionality. Having gone through four stages, which are conical / pleated paper cups, waxed paper cups, straight double-wall cup and laminated plastic coated paper cups, paper cup has gone beyond the pure function of drinking after so many years of development. According to the need of market and consumers, designers have designed a variety of creative paper cups, and paid more emphasis on the sense of design, environmental protection and convenience features, such as the easily identifiable cups, cups with handles, cups with stirs, folding cups and environmentally friendly paper cups. Next we will introduce to you the excellent cups designed by the world-renowned graphic designers.

随着社会经济的迅猛发展，纸杯的设计也是日新月异，不仅满足了人们的审美需要，也更多的注重纸杯的功能性。纸杯的发展经历了四个阶段，圆锥形/折褶纸杯、涂蜡纸杯、直壁双层杯和淋膜涂塑纸杯。经过这么多年的发展，纸杯已经超越了纯粹的喝水功能。设计师们根据市场和消费者的需要设计出了很多有创意功能的纸杯，更多的注重了纸杯的设计感、环保和方便功能。如易于识别的纸杯、带手柄的纸杯、带搅拌功能的纸杯、折叠式纸杯、环保纸杯等。接下来我们就将为你介绍世界著名平面设计师们的优秀纸杯设计作品。

Paper Cup with a Lid
带盖子的纸杯

This kind of cup is composed of a cup body and a lid. Compared with those cups without lids, the paper cups with lids could be re-closed and more effectively keep off the impurity. Specifically, this cup includes two forms: one is the paper cup with a visible lid and the other is the paper cup with an invisible lid. The cup body is made of frustum cone or cylindroids – like paper cups, which fits flawlessly with the transparent or opaque lid. This cup with simple structure features simplicity in structure, convenience in operation, novel appearance and low price. The simple closed structure could well preserve heat and also prevent from dust or mosquitoes to fall into; hence it can be regarded as a dynamic new product with simple structure and complete functions. The visible lid, the transparent material could present the inner beverage completely and thus can be used for containing ice cream and other cold drinks. The invisible lid, which is made of special material can withstand the hot water vapour and be used for containing hot drinks, such as coffee, hot fruit juice, etc.

该纸杯由纸杯体和纸杯盖组成。一般的纸杯没有盖，当不用的时候，往往会有杂质进入到纸杯中，当下一次用时不洁净。这种加盖纸杯在使用后将盖盖上，下次用时干净而且使用方便。

具体涉及一种配有可视性杯盖的纸杯和配有不可视杯盖的纸杯。由筒体为圆台或圆柱形的纸杯构成，其特征在于纸杯缘口与杯盖缘口相扣合，杯盖是由透明和不透明材质分别制成。这种纸杯结构简单、操作方便、区别显著、成本低廉，杯盖简单的扣合设计既具有保温的效果，又可以防止灰尘、蚊虫的落入，是一种结构简单、功能齐全非常有市场活力的新产品。可视纸杯，杯盖透明，可以看到里面的饮品，适合盛装冰淇淋、冷饮料。不可视纸杯，这种纸杯的杯盖有特殊材质制成，不会被水蒸气烫变形。所以这种纸杯适合盛装热饮，如咖啡、热果汁之类。

The picture above shows the Yogurt cups that are designed for Cone Kings Frozen Yogurt by Arjan van Woensel.
The picture below shows the ice cream paper cups that are designed for Orange cup by John Swieter. The transparent lids can not only present the inner ice cream, but give the product a sense of cleanness and coolness.

上图为加盖的酸奶纸杯，是由Arjan van Woensel为Cone Kings Frozen Yogurt所设计。
下图为John Swieter为Orange cup所设计的冰淇淋纸杯，透明盖子的设计不仅能让顾客清晰的看到冰欺凌，又使该产品看起来更加干净、清凉。

Paper Cup with a Handle
带手柄功能的纸杯

With the rapid development of economy, any best-selling products will quickly cause a large number of companies to flock into the same market. The identifiable differences between products are becoming increasingly blurred, which is also very prominent in the paper cup market. At this time we need to take into account the subtle differences of the product value, that is, to design the products based on the potential demand of consumers and carry out targeted marketing, so as to make it out of the homogeneity trap. Essentially, the functional design is the deepening of the market segmentation theory. There are various market segmentation techniques, all of which are actually based on functions.

Today, the commercial warfare has evolved into the psychological warfare of consumers. The winners are always those who firstly understand the customers' buying motives. In the functionally segmented market, it is not difficult to find new leading brands with absolute advantages. Designers who pay much attention to the products' function will play the important role in guiding the enterprise to be a famous brand. Also, the design of paper cup emphasises more on its functionality. Whether intentionally or unintentionally, the designer intends to select and use various materials to create a distinctive cup. An excellent designer should know how to use these materials to innovate and present the spirit of the times in his own design and resonate with the consumers.

Considering the thin sides of cup, a handle attached on it could ensure the hands not to be hurt by the hot drinks inside. This kind of cup not only benefits the users but also gives a sense of design. It is this small design that brings a lot of convenience for our lives. Using this cup in public places will surely attract much more attention. The cup printed with corporate logo will be a good practical advertment. This common daily necessity has got enough value of the overflow.

随着经济的不断发展，任何一种畅销的产品都会迅速导致大量企业蜂拥同一市场。产品之间的可识别差异变得越来越模糊，这种现象在纸杯市场中也极为突出。这时就需要我们考虑到产品使用价值的细微功能差异性，即以消费者的潜在需求为依据设计产品的功能，开展针对性的营销，使之跳出产品同质化陷阱进行设计。功能设计实质上是市场细分理论的深化，市场细分方法有好多种，但归根结底都是以功能细分的。

今天的商战已演变为消费心理战。战场的胜利者总是那些最早破译顾客购买行为动机的设计。在功能细分后的市场，往往能出现具有绝对优势的新领导品牌设计。功能设计师就是帮助企业成为名牌的指路人。纸杯的设计也是更加强调其功能性。无论是有意还是无意，设计师们都在想方设法的吸取来自四面八方的养料，在自己的纸杯设计中体现出来。好的设计师知道该如何运用这些养料，去超越、创新，在自己的设计中体现时代精神，并与消费者产生共鸣，令人们回味。

由于纸杯杯壁很薄，人们在喝热饮时，没有办法在很烫的时候拿起，这种防烫手柄纸杯就解决了这个问题。在饮用时可以拿着手柄，不仅方便实用，而且极具设计感。小小的设计给我们的生活带来很多的方便。我们在公共场所，用这种杯子喝饮品时，还会吸引很多人的眼球。如果在这种杯子上印上企业标志，这将是一个传播性极强的实用广告。这一司空见惯的日常用品，获得了极具张力的溢出价值。

The cup with a handle is designed by Vassiliki Argyropoulou for Draculi Coffee. The cardboard-based handle is stuck at the edge of cup which can effectively avoid hurting hands. However, this design is only suitable for small paper cups.

这是由Vassiliki Argyropoulou为Draculi Coffee所设计的带手柄的纸杯。该手柄采用硬质纸板卡在纸杯的边缘，这种纸杯的好处就是可以有效避免咖啡过热产生的烫手感觉，但这种设计只适合承载量不是很大的小巧的纸杯。

Paper Cup with a Stick
带搅拌功能的创意纸杯

In the current situation, our design needs creative ideas, for the reason that creativity can not only create wealth, but bring a new experience for our lives. The so-called creativity, whose basic meaning is the creative idea or a good idea that has never been known before. Of course, it is not unfounded, but re-designed on the basis of existing experience and materials. This design strategy features foresight, purposiveness, pertinence and utilitarian. Creativity comes along with working; working itself carries the creativity and creates properties innately. In the process of working, people get more and more creative consciousness and creative talent, thereby creating a rich culture.

Designers should pay more emphasis on creativity, according to the requirements of the market, consumers, considering bringing convience and surprise to people.

The cup with a stir bar is just the creative masterpiece that is produced according to the needs of consumers. When people drink coffee, tea and fruit juice, sometimes a stir bar will help making the drinks mix thoroughly. This kind of creative cup has greatly benefited the consumers.

在当前的形势下，我们的设计需要制造创意。因为创意不仅能够创造财富，还能为我们的生活带来不一样的新体验。所谓创意，它最基本的含义是指创造性的主意，一个好的点子，一个别人没有过的东西。当然，创意这个东西不是无中生有的，而是在已有的经验材料的基础上加以重新设计组合。这种设计策略，具有前瞻性、目的性、针对性、功利性的特点。创意是与人的劳动所共生，劳动本身就携带着先天的创意和创造属性。人类在漫长的劳动中，不断增长创意意识和创意才能，从而创造出丰富的文化。

设计师们更加应该注重创意，根据市场、消费者的要求，更多的考虑到是否会给人们带来方便、带来惊喜为出发点进行设计。

这种功能创意搅拌纸杯就是根据消费者的需要，重新整合设计的杰作。当人们在喝咖啡、茶和果汁时，很多时候都需要一个搅拌棒，使之饮品调匀。当我们在为这个发愁时，这种功能创意搅拌纸杯的诞生，方便了消费者更好的享用饮品。

The paper cup with a stick is designed by Phuong Ngoc Le. This cup functions just like an instant noodles cup: consumers have the choice to enjoy the fresh fruit first and add hot water for a cup of tea later. A fork/spoon would be provided as a handy tool to serve the fruits.

这是由Phuong Ngoc Le设计的带搅拌功能的纸杯。这一纸杯的功能犹如一个即食方便面纸碗：消费者可以选择先品尝新鲜的水果，随后在杯子里加入热水对茶叶进行冲调。与此同时，设计师还精心地在杯中准备了一个叉/勺，以方便消费者品尝水果。

Paper Cup with a Sleeve
带杯套的纸杯

Cups have become necessities in our lives. Along with the improvement of the type and design of paper cups, the requirements for design are also increasing. The utility function of paper cups is becoming more and more important. Hot drinks in paper cups will be very hot, so there is no way to directly pick up the cup. Paper cup with a sleeve is just like a cup wearing a coat, which is more beautiful, practical, and suitable for this fast-paced lifestyle. You don't need to wait; just hold it directly to go to work or shopping.

You can also print on your own brand logo, highlighting the product brand features in the cup. In view of choosing materials for sleeves, on one hand, you should choose anti-hot insulation materials, and on the other hand, you should take into account the cylindrical form and skidproof materials. Additionally, other elements such as the printing process, texture, and quality also cannot be ignored.

This kind of paper cups has seamlessly combined publicity and practicality. In regard of the effect of advertising, its advantages and effects of publicity could exceed other advertising media.

纸杯在人们生活中经常出现，它已经成为我们生活中的必需品。纸杯的种类和设计也是日益增多，对纸杯设计的要求也增多了。纸杯的实用功能也越来越受到重视，当人们在用纸杯喝热饮时，会很烫，没有办法直接拿起杯子。这种带杯套的纸杯，给人们感觉像是为纸杯穿上了外衣，更加美观、实用，适合我们现在这种快节奏的生活方式。不需要等待，可以直接拿着它去上班、逛街。

还可以在杯套上印上自己品牌的标志，突出产品的品牌特征，为品牌做宣传。在杯套的材料选择方面，一方面要选用防烫隔热的材料，还要考虑到杯体是圆柱形的，要选择防滑的材料。除了这些基本的因素外，还要考虑到最终效果，印刷工艺，纸张的手感，是否容易掉毛掉粉等因素。

这种纸杯将实用性和杯体的广告性合二为一。在广告的有效传达上，具有其他广告媒介难以比拟的优势和效果。

The above cups are designed for One Tree Coffee. The sleeve is made of brown paper, whose design not only highlights the brand's identity, but benefits people to pick up the paper cups to drink hot coffee.

The below cup is designed for Wi-Fi Coffee Bar. The sleeves take the logo of the bar as the design elements, highlighting the characteristic of the brand.

上图是One tree Coffee的咖啡杯，该纸杯的杯套采用牛皮纸包裹，这种设计不仅突出了该品牌的标识，还方便了人们在喝热咖啡时拿起纸杯。

下图是Wi-Fi Coffee Bar的纸杯，该纸杯的杯套采用的元素是该咖啡吧的标识，突出了产品的品牌特征。

Folding Paper Cup
折叠式纸杯

Folding disposable paper cups or strips aim to save space and keep clean as well as easy to take away. Its features are the following: it is composed of side wall and bottom plate; the side wall of a cup is in a straight barrel shape; the side wall is evenly distributed with four axial polylines; two folding lines correspond to four folding lines at the bottom plate; the bottom plate is in a spherical shape; the centre of bottom plate is lower than the junction part of bottom plate and side wall while the centre of bottom plate is higher than the edge of the bottom of the side wall.

The foldable disposable paper cup could save a lot of space in the process of taking away or storing or even discarding, which benefits the user greatly; disposable paper cup strips are easy to keep clean when taking along and in storage.

The foldable paper cup is made up with the cup body and shaking cover, and the cup body includes the sides of left, right, bottom, front and back. It features the diagonal straight folding lines pressed on the sides of bottom, front, left and right; the folding lines pressed on the sides of bottom, front, left and right appear in an oblique state after the body is unfolded. The beneficial effects of this paper cups are: minimising the volume of paper cups and benefiting to take away; simplifing the using process; ensuring good impermeablity; increasing the commercial value, for the reason that it can be packed together with the tea bags and other beverage package; can be reused as a small storage box or flowerpot; the surface of the cup can be treated with a variety of decorative techniques, increasing the product with a sense of fashion.

可折叠式一次性纸杯及一次性纸杯条，旨在提供一种占据空间小、不易弄脏且便于携带的一次性纸杯。其特征在于，一次性纸杯由侧壁和底板组成，侧壁为直桶状，且侧壁上均匀分布有四条轴向折线，其底板上设有与四条折线相对应的两条折线，底板呈球面状，其底面中心处低于底板与侧壁连接处，且底面中心处高于杯体侧壁底部的边缘。

可折叠式一次性纸杯在携带、存储或丢弃时节省大量的空间，为使用者提供很大的便利；一次性纸杯条在携带和储藏的过程中不易使杯子被弄脏保持卫生。

折叠式纸杯，包括可折叠的杯体和摇盖，杯体包括左面、右面、底面、正面、背面，上面分别压制有对角的直线折痕线，折痕线在杯体展开后是同一斜向状态。极大限度的减小了纸杯携带时的体积，方便用户随身携带使用；使用步骤简单方便；折叠后的纸杯具有良好的密闭性，保证其卫生安全；纸杯可与茶包等饮料包一起包装出售，增加商业价值；纸杯使用后可废物利用，作为小花盆、小储物盒，有利于环境保护；纸材表面可有多种装饰手法，增加产品时尚感。

HELP those in need

Concept

The hand as a profound effect, not only communicative but the join of hands to pray, to ask support and give something. The join of hands to ask, which forms the cup design, helps to mark the year 2010 as European Year of Combating Poverty and Social Exclusion. The cup has no support base, as people who need support and it´s safe by the information brochure.

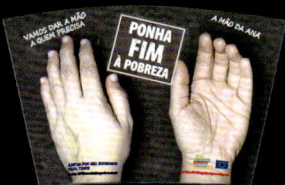

This is designed by Ana Silvia Santos for 2010 European Year of Combating Poverty and Social Exclusion. The designers have taken the rare folding technique to design the bottom of the paper cup and customly designed a special cup holder in view of the entire awl-shape cup.

这是Ana Silvia Santos为"2010 European Year of Combating Poverty and Social Exclusion"所做的纸杯设计。纸杯的底端采用了极为少见的折叠式设计，由于整个纸杯的形状是锥形，所以设计师又给该纸杯配了专门的杯托。

Anti-hot and Anti-slip Paper Cup
防烫防滑纸杯

The anti-hot and anti-slip paper cups are composed of a bottom and a cylindrical side wall, both of which are made up by gluing single layer paper. The outer surface of the side wall is scattered with many protruding bumps. Compared with ordinary cups, this cup can significantly reduce material consumption and the complexity of the manufacturing, thereby reducing the manufacturing costs.

In an age when high quality of life is required, apart from the appearance of paper cups, the textures of them being in hands are also important. The anti-hot and anti-slip paper cups have met this requirement. However, its appearance is slightly inferior, for it can be only printed with the logo and some simple pattern graphics.

Paper cups play multiple roles in people's lives, and the changing requirements of the cup design encourage designing perfect cups.

防烫防滑纸杯，包括有用单层纸粘接成型的杯底和筒形杯壁，杯壁外侧面上分布有许多向外凸起的凸点。这种纸杯与现有普通纸杯相比具有防滑落的效果，与现有防烫纸杯相比可以大大降低材料消耗，降低制造设备的复杂程度，从而降低制造成本。

人们在追求高品质的生活时，不仅对纸杯的外表设计有要求，更对纸杯拿在手里的质感提出了新的要求。防烫防滑符合了这种质感的要求。但是在外观设计上也略显逊色，只能印上标志及一些简单的图案图形。

纸杯在人们生活中扮演多重角色，对纸杯设计要求的变化，成为能设计出更加完美的纸杯的催化剂。

This cup is designed for Le Charme brand. It employed special materials and technology to have the surface be specially treated, so as to protect your hands even if the water temperature rises up to 100 degrees.

这是为Le Charme品牌设计的纸杯。该纸杯采用特殊材料和工艺做了纸杯表面的处理，即使是水温高达100℃，也不会感觉烫手。

Recognisable Paper Cup
易于识别的纸杯

Disposable cups because of its low price and easiness to use have become widely used in people's daily life, especially in public places. Disposable cups are mass-produced, and the appearance and size of the cups are similar, so people often take others' cups by mistake, not only unsanitary, but a waste of resources.

The user-friendly design will become the dominant direction of design in the future, which focuses more on human behaviour habits, the body's physiological structure, the human psychology, the way people think and so on. Additionally, it fully reflects the respect for the needs of human psychology and physiology as well as the spirit. This humane design is a combination of science, art and humanity, enriching the product with a sense of beauty and fun.

A small detail will benefit our lives. In order to solve this problem, designers have designed this easily identifiable paper cup, which is composed of letters, numbers, the combination of letters and numbers or even simple graphics. The beneficial effects of such paper cups are: avoiding being reused, reducing health risks caused by cross using and preventing resources waste. Now, this kind of paper cups is used extensively in public places.

一次性纸杯因其价格低廉、使用方便的特点，成为了人们日常生活中广泛使用的必需品，尤其在公共场所使用量很大。但由于一次性纸杯是批量生产的，纸杯的外观和尺寸都是相同的，因此人们在使用中经常会出现错拿别人纸杯的问题，不仅不卫生，同时造成资源浪费。

人性化的设计将成为未来设计的主导方向，在设计的过程中，注重人的行为习惯、人体的生理结构、人的心理情况、人的思维方式等方面。并在设计中体现对人的心理生理需要和精神追求的尊重和满足。这种人性化的设计是科学与艺术、艺术与人性的结合，并使这种艺术和人性富于美感，充满情趣和活力。

一个小小的细节设计，为我们的生活带来方便。为了解决这样的问题，设计师们利用字母、数字或简单的图形，设计了这种易于识别的纸杯。这种纸杯的有益效果是：避免其他人再次使用，减少了交叉使用的卫生隐患，避免浪费资源。现在这种纸杯被大量的使用在公共场所。

The ABC cups were designed by Sunhan Kwon, with a set of 24 cups, each of which is printed with a letter. Just remember the letter; even among many people, it will not be mistaken.

该ABC纸杯是由Sunhan Kwon所设计的，这套纸杯总共24个，每个纸杯上印有一个英文字母。只要记住了自己的字母，即使在人多时使用，也不会因为混淆拿错了纸杯。

Eco-friendly Paper Cup
环保的纸杯

Along with the great developments of society, the standard of human life as well as the production efficiency, the issuses of energy shortage and industrial wastes are increasingly highlighted, which requires the products about human lives and works should be produced in a simple and clear way, with a sense of modern associating with the modern information age and having a spirit to correspond with the modern life. Green, environmental protection concept has become the mainstream of today's society. Seen from the history, the direct impacts of green design have been referred by the design theorist Victor Babang, who published a book entitled "Design for the Real World" in the 1960s, emphasising that the design should seriously consider the use of the Earth's resources and protect the Earth's environment. Since then, green design has also got more and more attention and recognition.

Green design means that in the design of the product and its life cycle, its impact on resources and the environment should be taken into account. Apart from the product's function, quality, development cycle and cost, all relevant factors should be optimised in order to minimise the negative impact of the products and its manufacturing process on the environment and meet the stanard of environment protection. In the whole process of green design, the products' recyclability, maintainability, reusability, etc. are the foucuses and aims of the design. What's more, the green material selection should be taken into account.

The green materials are mainly from the natural plant fibre which will soon rot, do not pollute the environment and can be recycled to produce paper. Many large international companies use recyclable paper to design the annual report, publicity materials and stationery to reflect their concerns for the environment, establishing a good corporate image. Such environmentally friendly paper cups will replace normal cups and pour into our lives.

时代在前进，人类生活水准在提高，生活节奏在加快，生产效率在突飞猛进，但同时面临能源的短缺、工业垃圾日益增加等诸多困惑。这些就要求伴随人类生活、工作的产品应该简洁明快，新颖亲切，具有一种与信息时代相关联的现代感，包涵一种同现代生活相符合的精神。

绿色、环保已经成为当今社会的主流。从历史可以看出，对于绿色设计产生直接影响的是美国设计理论家维克多·巴巴纳克，在 20 世纪 60 年代他出版了一本名为《为真实世界而设计》的书，强调设计应该认真考虑地球资源的使用问题，并为保护地球的环境服务。从此绿色设计也得到了越来越多的人的关注和认同。

绿色设计是指在产品及其寿命周期全过程的设计中，要充分考虑对资源和环境的影响，在充分考虑产品的功能、质量、开发周期和成本的同时，更要优化各种相关因素，使产品及其制造过程中对环境的总体负影响减到最小，使产品的各项指标符合绿色环保的要求。在整个绿色设计过程中，着重考虑产品的可回收性、可维护性、可重复利用性等，并将其作为设计目标。纸杯的设计也是与时俱进的，更多的考虑到纸杯的环保材料选用。

这种环保纸杯的原料主要是天然植物纤维，在自然界会很快腐烂，不会造成环境污染，也可回收重新造纸。因此，许多国际大公司使用可回收纸用于年报、宣传品制作，用回收纸制成信笺、信纸以体现其关注环境的绿色宗旨，同时又树立了良好的企业形象。这种环保纸杯将会取代正常纸杯，大量进入我们生活中。

100%
COMPOSTABLE
USES 65% **LESS** CO_2
NO PETROLEUM
NO SLEEVE REQUIRED

WELCOME TO A CLEANER WORLD
repurposecompostables.com

The eco-friendly paper cups were designed by Vera Valentine, advocating cherishing the Earth and caring for trees.

该环保纸杯是由Vera Valentine所设计的，倡导爱护地球，爱护树木。

Materials of Paper Cup
纸杯的选材

Sometimes, when you see a paper cup or use a paper cup to drink water, you will feel the material of the cup body. And the choice of cups' materials is related to people's health, so people pay much attention to the choice of materials. Usually talking about raw materials, disposable cups are divided into two types: cold drink cups and cups for hot drinks.

On the market, materials for paper cups have become increasingly diverse. The common type of white cardboard, mainly used to hold dry things cannot be filled with water and oil; secondly, the waxed paper cups are more waterproof and thicker, but as long as the water temperature exceeds 40°C, the wax will melt; thirdly, now widely used are paper and plastic cups with a layer of paper outside and a coated layer of paper inside; if the material is poor and the processing technology is not good, it will also have a harmful substance. With the increasing concerns of people's health, designers have considered the importance of environmental issues to bring a good impact on the client's choice of materials.

有时候，当你看到一个纸杯或者用纸杯喝水时，你会感受它的杯体的材质，并且纸杯材料的选择关系到人们的身体健康，所以人们很重视纸杯材料的选择。一次性纸杯原材料通常来讲分为两种：冷饮杯和热饮杯。

目前市场上的能做纸杯的材料越来越丰富，常见的一种是用白卡纸做的，主要用来装干东西，不能盛水和油；第二种是涂蜡纸杯，这种杯子因为用蜡浸泡过，所以较为防水、厚实，但只要杯子里所装水的温度超过 40℃，蜡就会融化；第三种是现在人们普遍使用的纸塑杯，外面是一层纸，里面则是一层淋膜纸，如果制作时所选用的材质不好或加工工艺不过关，也会产生有害物质。随着人们对健康关注度的提高，设计师也把环境问题的重要性作为自己份内的事，并尽量对客户的材料选择造成好的影响。

Cups for Cold Drinks
冷饮杯

Cups for Cold Drinks: They are generally used to hold floating ice drinks. The cup body must be moisture-proof, usually taking the wet and dry wax as the main materials for the inner wall.

冷饮杯：一般用来装浮着冰块饮料。所以杯身设计必须防潮，通常选用湿蜡和干蜡做为纸杯的内壁。

They are designed by Kelvin Ng for Soyato ice cream, which advocates guiding a healthy and fun lifestyle.

这是由Kelvin Ng所设计的Soyato冰淇淋纸杯,该产品提倡引导健康、充满乐趣的生活方式。

Cups for Hot Drinks
热饮杯

Cups for Hot Drinks: The inner surface of paper cup with a layer of plastic. It is usually used for containing coffee and hot juice. In the process of designing, the stiffness and basis weight of paper should be considered in order to protect the users.

热饮杯：纸杯内壁有一层塑料。通常装咖啡、热果珍等。在设计时特别考虑纸张的挺度，要选择定量高一些的纸张，以保证饮用者安全。

These cups were designed by Sergio Laskin for Coffee Time. The cup is made of a safe and environmentally friendly material, giving a sense of warmth and uniqueness.

这是由Sergio Laskin为Coffee Time所设计的咖啡纸杯，该纸杯采用安全环保的材料制成，色彩感觉上也体现了该咖啡吧温暖、别致的特征。

This set of cups has chosen the most typical coffee colour. The graphic design shows the uniformity through large, medium and small cups, while each of the cups has its own characteristics. The designer also designs a take-away coffee handbag exclusive for this coffee shop, the handbag modelling in the brand Logo "big beard" makes it really impressive and interesting. This handbag uses eco-friendly corrugated paper and is very convenient to carry.

这组杯子选用了最具有代表性的咖啡色。大中小三种杯型上的图案也选用了统一的设计，但每个又有各自的特点。设计师还专门为该咖啡店设计了外卖咖啡的手提袋，手提袋的造型则是该品牌的LOGO"大胡子"，显得很有特色也非常有趣。该手提袋采用环保的瓦楞纸设计，方便人们携带。

Design Elements of Paper Cup
纸杯的设计元素

Apart from the function, the design of paper cups pays more attention to conveying an attitude of life and a brand concept to the consumers. Therefore, throughout the design process, the visual appearance of surface should not be ignored. This refers to the design elements that each designer is familiar with: graphics, text, patterns, illustrations, trademarks, company names, colour, etc., all of which should be combined by means of clever design.

The key of cup design lies in finding and creating, which needs the designers' continuous developing and experiencing. To impress the cusumers is a challenge for the designers. Creating an impressive cup, it is necessary to pay more attention to the details of design, developing creative graphics, colour composition and material texture as well as the artly organic combination of a variety of elements. Also, designers should be aware that a strict attitude could truly move the viewers.

Design is the combination of designer's imagination and the expression of this imagination. In this process of imagination, varied design elements and all of the design methods should be employed. The design elements of the paper cup can be divided in respect of the functional structure or the form of cup patterns. We have previously introduced functional cups, and next we will introduce you the design of the cup.

纸杯的设计除了满足功能上的需求，更重要的是向广大的消费者转达一种生活态度，一种品牌理念。因此，在整个设计过程中，不单单注重功能性，更注重杯体表面视觉上的美观。这就用到了对于每个设计师都不陌生的设计元素：图形、文字、图案、插图、商标、公司名称、色彩等，通过巧妙设计和组合。

纸杯设计的关键之处在于发现，只有不断通过深入的感受和体验才能做到，打动别人对于设计师来说是一种挑战。纸杯设计要让人感动，就要在细节设计上更加留心，图形创意本身能打动人，色彩品位能打动人，材料质地能打动人……把设计的多种元素进行有机艺术化组合。还有，设计师更应该明白严谨的态度自身更能引起人们心灵的振动。

设计是设计师的想象及对这种想象的表达的结合，在这个想象的过程中可以使用丰富的设计元素或者一切设计手段。纸杯设计中的元素可以从功能结构划分，又可以从杯体图案形式上划分。之前我们已经介绍了功能类纸杯，接下来将为你介绍杯体的设计。

Logo Paper Cup
企业标志纸杯

When the world entered the brand competition age, the importance of the corporate logo on the paper cups is particularly highlighted. Logo is a mark to show the characteristic. It uses simple and recognisable images, graphics or symbols to show the spirit, philosophy, culture and personality. As a special mode of the intuitive contact, it not only goes in social activities and production activities, but plays an important role in the fundamental interests of the country, social groups and even individuals. This asks for designers to create corporate logos or corporate symbols as bright as possible.

In general, this task can be completed in one of the simplest ways, which refers to directly printing the logo onto the cups. A brand with unique personality and values, could be reflected by some special effects technology, such as relief, embossing, hot paint and so on. Addtionally, to create a clever and interesting relationship between the brand name and the shape of cup body is also advisable.

Corporate logo on the cups often creates a tension between logo size and aesthetic concept, causing conflicts between the visual impact of the brand and the overall balance of the paper cup design. Some people think that the colours of logo should be the most striking and eye-catching; however, the corporate logo is not just a logo or symbol; it also involves a range of issues, such as the symbolic meaning, colours matching, font combination and styles, etc.

These effects and initiatives can cleverly and clearly convey certain information: corporate image, corporate content, etc. The paper cups printed with company logo can be taken to everywhere, which potentially brings a great advertising effect.

当世界进入品牌竞争的时代，纸杯表面的企业标志的重要性也尤为突出。标志是表明特征的记号。它以单纯、显著、易识别的物象、图形或文字符号直观语言，表现企业的精神、理念、文化和个性。标志，作为人类直观联系的特殊方式，不但在社会活动与生产活动中无处不在，而且对于国家、社会集团乃至个人的根本利益，越来越显示其极重要的独特功用。 这就对设计师提出了新的要求，就是让企业标志或者企业符号以尽可能耀眼的方式呈现出来。

一般来说，这个任务是可以用一种最简单的方式来完成的，就是直接把标志直接印刷在纸杯的杯体上面。一种品牌的个性和价值，也可以通过一些特殊的效果工艺传达体现出来，如浮雕、压花、烫漆等。还可以通过品牌的名字与杯体的外形之间形成一种巧妙有趣的关系。

企业标志在纸杯上的应用，常常在标志的大小和审美感觉之间造成一种紧张的关系，品牌的视觉冲击力和纸杯设计的整体平衡感之间会出现冲突。有人认为"标志的颜色越醒目、越大越好"，然而，企业标志不仅仅是一个标识或符号，它还牵涉到象征寓意、颜色搭配、字体组合、风格特征等一系列问题。想要宣传企业，并非一个标志就能覆盖的。

这些效果和举措，可以巧妙明确地传达出某些信息：企业形象、企业内涵等。纸杯的大量使用，使得印有企业标志的纸杯，被人们拿着穿梭在大街小巷，无形中带来了巨大的广告效应。

Patterned Paper Cup
图案纸杯

As the name suggests, pattern features with a sense of decoration as well as an orderly and shapely structure. Similarly, pattern design focuses on the beauty sense of order and decoration, varying in the unification. Pattern is being widely used, such as in the design of graphics, advertising, costume, packaging and even interior space. The role of pattern design is also very important, for it not only beautifies our environment, but makes life more magic and colourful.

Of course, design of paper cups cannot live without patterns. According to the customers' requirements and understanding of the enterprise, designers have employed image-symbol to design patterns to meet the aesthetic need of people. The main design elements composing a pattern often include point, line and surface. According to the point, line, surface and visual psychology of colour, designers make good use of the principles of contrast and unity, rhythm and cadence, proportion and balance, orderliness and repetition to combine the patterns with the cup body seamlessly.

Patterns on the paper cups are to cause positive interaction between the consumer and the cups, enhancing corporate brand image. No matter what kind of pattern is used in a paper cup, its power is derived from the designer's own inherent ability, which needs to simply and rapidly convey the information to the universal. Therefore, the design of the pattern paper cups is to condense the design concept into the simplest pattern. Good patterns can exceed all words.

Patterns on the cup body can not only improve people's appreciation ability of beauty, but create beauty in practical applications, making people enjoy drink and the beauty at the same time.

图案，顾名思义就是有装饰意味的、结构整齐匀称的花纹或图形。图案的设计具有秩序的美感，也就是排列整齐，装饰性强，统一之中又有变化。图案的应用领域相当广泛，比如平面设计、广告设计、服装设计、包装设计以及室内空间设计中。同时，图案设计的作用也是非常重要的，它不仅美化了我们的环境，而且使生活变得更加幻丽多彩。

纸杯的设计当然也离不了对图案的应用，设计师根据客户要求和对企业的了解，运用图象符号设计出符合人类审美的图案。构成图案设计的要素主要是点、线、面。根据点、线、面以及色彩的视觉心理，运用对比与统一、节奏与韵律、比例与权衡、对称与平衡、条理与重复等形式美的原则，与纸杯的杯体巧妙结合。

纸杯上的图案设计，是为了让消费者与纸杯之间在情感与体验上获得良性互动，提升企业品牌形象。无论在纸杯上使用什么样的图案，他们的力量都是源于设计师自身内在的能力，这就是要简单快速地传递信息，而且具有普遍有效性。因此，图案纸杯的设计就是一种把设计理念凝结到最简单的图案中去。好的图案可以超越一切文字阐述。

纸杯杯体上的图案，不仅能提高对美的欣赏能力，而且还能在实际应用中创造美，使人们在享用饮品的同时也得到美的享受。

Illustrative Paper Cup
插画纸杯

Illustration is a common language around the world. It features images and tries to make lines and forms be clear and simple according to the unity of aesthetic concept and practice. In view of the commercial application, it includes images of people, animals and products.

Illustration is also the basic design approach in the design of paper cup, while with the development of photography, illustration is used less and less on the cups for its a little sense of hand-made and tradition or the changes of designers' personal preference. However, with the development of science and technology, this issue won't limit the application of illustrations.

The illustration is inseparable with the paper cup design, for the reason that it can enable us to better understand the business and add fun for the paper cups.

Illustration could provide a variety of styles and forms for the cups. This means that the role illustration plays in conveying the brand's core concept, the designer's design philosophy, and highlighting the brand personality aspects is no less than photos.

Of course, in some cases, photos can be better to achieve a certain effect, but due to the limitations of the paper cup body, it is not as good as illustration. Illustration can be modern, interesting, stylish, charming and humorous... To design more innovative and creative illustration paper cups requires more for the designers and asks for the combination of designers and illustrators. Only in this way can produce the paper cup with a sense of life.

插图是世界都能通用的语言。插画是运用图案表现的形象，本着审美与实用相统一的原则，尽量使线条、形态清晰明快，制作方便。其设计在商业应用上通常分为人物、动物和商品形象。

插画也是纸杯设计的基本常用方法，但是随着摄影技术的发展，插画在纸杯上的应用越来越少。也可能是因为插画中带有一些手工制造和传统的色彩，而在我们这个高科技的现代社会里，这些观念又显得不合时宜。或者也是因为设计师个人偏好发生了变化。随着科技的发展，这个问题已经不能成为制约插画应用的原因。

虽然有这些因素，但是插画设计仍然与纸杯设计密不可分。因为插画可以使我们更好的了解这个企业，也为我们的纸杯设计增添趣味性。

插画能为我们的纸杯设计提供多样化的风格形式。这也就意味着，纸杯在传达品牌的核心观念、设计师的设计理念以及凸显品牌个性等方面，插图的作用不亚于照片。

当然，在有些时候，照片是可以更好的达到某种效果，但是由于纸杯杯体的局限性，在这点上就不如插画所显示的那样。插画可以是现代的、有趣的、时尚的、迷人的、幽默的……要想设计出更多新颖有创意的插画纸杯，这就对设计师提出了新的要求，设计师和插画家相融合，才能设计出富有生命力的纸杯。

Words Paper Cup
文字纸杯

Words are important parts of human culture. No matter in what kind of visual design, words and images are indispensable. The quality of typography directly affects its visual communication. Therefore, the application of words is an important design element to enhance the visual communication effect, improve the infectivity of work and give the design a unique aesthetic value.

Of course, the design of cups also needs words. The words' application should be subordinated to the style of the cups. The font selection, layout and design differ with various brands. In order to make the brands more distinctive and unique, many designers tend to invent new fonts, renovate existing fonts or combine words with graphics. Since the words have the task of conveying the feelings of designers, they must have a strong visual sense of beauty. Now, to measure things with beauty and ugliness from the visual senses has become a standard consciously or unconsciously, so to meet the aesthetic needs of people and improve their taste is the task of every designer.

Apart from the words design, the relationship between words and the whole cup as well as other design elements also should be considered, which requires designers to pay more attention to detail design.

Each cup has its own unique style. In view of this, on a paper cup, various words combination as well as the combination of text and graphics should go well with the entire paper cup design style and convey a sense of wholeness, but each element shouldn't be in individual style respectively. Words paper cups are also in line with the current social emphasis of corporate culture.

文字是人类文化的重要组成部分。无论在何种视觉设计中，文字和图片都是其两大构成要素。文字排列组合的好坏直接影响其视觉传达效果。因此，文字的应用是增强视觉传达效果，提高作品的诉求力，赋予设计审美价值的一种重要设计元素。

当然，纸杯设计的设计也需要文字。纸杯中文字应用要服从于纸杯的设计风格特征。纸杯中字体选择、编排、设计，受到品牌的差异的影响。为了使自己的品牌更具特色、与众不同，很多设计师采取发明新字体、对已有字体进行改造、文字与图形的结合等。设计师运用这些文字传达出有感情的设计，因而它必须具有视觉上的美感，能够给人以美的感受。人们对于作用其视觉感官的事物以美丑来衡量，已经成为有意识或无意识的标准。满足人们的审美需求和提高美的品位是每一个设计师的责任。

纸杯设计中的文字应用，不仅仅是对文字本身有要求，更要求文字不仅仅体现在局部，而要注重文字应用于纸杯中的整体感，以及文字和色彩的搭配。这就要求设计师对细节设计更加重视。

对于每个纸杯设计而言，都有其特有的风格。在这个前提下，纸杯上各种不同文字的组合、文字和图形的组合，一定要具有一种符合整个纸杯设计风格的倾向，形成总体的情调和感情倾向，不能文字、图形、标志各自成一种风格，各行其是。这种文字纸杯也更符合了目前社会对于企业文化的重视。

Graphics Paper Cup
图形纸杯

Graphic design develops and enriches itslef with the improvement of modern design, which not only reflects people's acquaintance of graphics, but illustrates its particularity and prescriptiveness as a social and historical phenomenon. In the information age, the spread of graphics has become more and more important in the human lifestyle. In the business community, modern graphic design is omnipresent. All kinds of information it conveys can be easily understood and used by people from different cultural backgrounds and countries.

Graphics are the commonly used design element in the design for its features of visualisation, powerfulness and endurance. It can concentrate on the core philosophy and corporate culture and convey the information quickly and effectively. Scientists have shown that graphics are easier to be remembered than words.

Addtionally, graphics are always used in the design of paper cups. No matter it is colourful, or black and white, it can more intuitively embody the philosophy of the company or enterprise. It could simply make a paper cup distinguish from others and stand out.

What's more, the graphics could help customers select their preferable paper cup among so many kinds, since it can clearly reveal the characteristics of the brand and convey the brand's value, style and aim.

This kind of paper cup features graphics and emphasises employing simple approaches to explain more problems.

图形设计伴随着现代设计的发展而不断丰富其内涵，这既反映了人们对图形性质的认识，也说明了图形作为一种社会历史现象的特殊性、规定性。信息化时代，图形的传播在人类生活方式中越来越重要，在商业社会中，现代图形设计无处不在。因为它传递的各种信息，是具有不同文化背景、不同国家语言的人都容易接受和使用。

图形是众多设计中常用的设计元素，因为它直观、有力并且持久。图形能将一个核心理念、企业文化等进行浓缩，并能迅速而有效地传达信息。科学家研究表明，图形更能被记住，并且记忆速度比文字快四倍。这个效果就归因于图形的独特性，因此，它能比较容易地被记起。

纸杯设计中也常常运用图形的设计。这种图形设计无论是彩色的、黑白的，都能更直观的体现这个公司或企业的理念。这种图形使各样纸杯都具有不同于别的纸杯的特点，这就有助于这种品牌的纸杯从另一品牌中脱颖而出，提升品牌形象。

当人们面对多种纸杯选择时，纸杯上的图形对纸杯的帮助起到很大的作用。这就归功于图形的设计，因为它有助于揭示品牌的特点，传达它的价值、风格与追求。

这种图形纸杯，其特点就是用图说话，简单的手法，说明更多的问题。

Imbibo Café

Production date: April 2011
Designer: Diana Pesce
Photography: Nathan Pesce
Nationality: American

Imbibo is Latin for drinking or to think up. The design was inspired by European architecture and Bauhaus designs, as well as, old American and European signage. The personality of the design is energising, bold, and sleek. This is communicated through a pairing of two modern san serif typefaces that touch on the typographic style of the past. The colour palette of rich red/salmon and dark brown were used to evoke feelings of warmth and bold flavour, which are parallel to the characteristics of the café's hot beverages.

"Imbibo" 咖啡厅

完成时间：2011年4月
设计师：戴安娜·派斯
摄影师：南森·派斯
国家：美国

"Imbibo"在拉丁语中寓意啜饮或发明。该设计方案的设计灵感源自欧式建筑、包豪斯设计作品以及欧美老式标牌的启发。这一系列咖啡杯的特色是活力无限、大胆前卫、精致细腻。两种现代无衬线字体在彰显出作品特色的同时，自然地激发观者对以往排版风格的思考。除此之外，大红色、鲜肉色与深棕色的完美搭配巧妙地营造出温暖、前卫之感，从而与该咖啡厅中热饮料的独特风味相得益彰。

The Europa Café

Design agency: Camila Drozd
Production date: April 2011
Designer: Camila Drozd
Photography: Camila Drozd
Client: Kenneth and Lidiya Dobosh
Nationality: American

The Europa Café is a charming little shop located on Main Street in Stroudsburg. It offers a wide range of delicacies from across Europe. The designer redesigned the logo to better reflect the style and feel of the café and offered the design to the owners free-of-charge. Although they loved the logo, Ken and Lidiya decided not to take the designer's offer due to the potential expense of changing everything. The offer is still on the table if ever they change their minds. (The designs simulate what the designer hopes the café sign and packaging will look like in the future.)

欧罗巴咖啡厅

设计机构：卡米拉·德罗佐德设计工作室
完成时间：2011年4月
设计师：卡米拉·德罗佐德
摄影师：卡米拉·德罗佐德
客户：肯尼斯，莉迪亚·都博什
国家：美国

欧罗巴咖啡厅是一个小巧而迷人的咖啡品尝空间，坐落在斯特劳斯堡的大街之上。该咖啡厅能够为客户提供全欧洲的各种风味美食。来自卡米拉·德罗佐德设计工作室的设计师卡米拉·德罗佐德自愿为该咖啡厅的标识进行重新设计，力图更好地展现出该咖啡厅的风格与个性，并向店主提供免费的设计服务。尽管店主肯尼斯与莉迪亚·都博什对设计师所设计的标识十分欣赏，但考虑到更换标识所产生的潜在花销，于是委婉地谢绝了设计师。然而，如果店主改变主意，这一设计方案还是存在很大的实施可能性（设计师所设计的这一方案完美地勾勒出其对未来咖啡厅标识和包装设计的憧憬）。

044

Royalt Café

Design agency: Thing Studio
Production date: 2010
Creative director: Carrie Lau
Designer: Carrie Lau
Photography: Ricky Chan
Client: Royalt
Nationality: American

皇室咖啡厅

设计机构：主题设计工作室
完成时间：2010年
创意总监：卡利·刘
设计师：卡利·刘
摄影师：里基·陈
客户：皇室咖啡厅
国家：美国

The purpose of this project was one of the pitching / proposed projects that Thing Studio aimed for Royalt Café. Royalt is a playful collision of space and it creates a cosy environment for customers to enjoy the fusion food and visual creation. The bright colour and fun graphic elements deliver the characteristics of the café. The project included logo redesign, paper cup, napkin and paper bag.

该项目是主题设计工作室向皇室咖啡厅提供的策划方案之一。皇室咖啡厅空间设计风格独特，拥有着极强的视觉冲击力，顾客在这一舒适的环境中品尝可口的融合美食的同时还能够尽情地享受到视觉上的盛宴。对于咖啡杯的设计，明亮的色调和妙趣横生的平面元素完美地传达出该咖啡厅的独有魅力。该项目包括标识的重新设计以及纸杯、餐巾纸和纸袋的设计。

Bite

Production date: 2010
Designer: Hannah Jackson
Photography: Hannah Jackson
Nationality: American

Branding and packaging design for a sustainably-driven bakery opening up in Leeds City Centre.

"咬一口"纸杯

完成时间：2010年
设计师：汉娜·杰克逊
摄影师：汉娜·杰克逊
国家：美国

这一项目是设计师汉娜·杰克逊专为一个以可持续理念为出发点的烘焙坊而打造的品牌塑造与包装设计方案，该烘焙坊坐落在利兹城的市中心。

InsTnt

Production date: 2011
Designer: Phuong Ngoc Le
Photography: Phuong Ngoc Le
Client: T2
Nationality: Australian

InsTnt is a sub-brand of T2 (Tea Too). Its name is the short form of ins-tea-nt, taking a twist from the original name of the brand. It is created to introduce fruits into the core brand as fruit is a perfect companion for tea. This cup functions just like an instant noodles cup: consumers have the choice to enjoy the fresh fruit first and add hot water for a cup of tea later. A fork/spoon would be provided as a handy tool to serve the fruits.

"InsTnt" 纸杯

完成时间：2011年
设计师：乐氏玉芳
摄影师：乐氏玉芳
客户：T2茶行
国家：澳大利亚

"InsTnt"是T2茶行的一个子品牌。"InsTnt"这个名字源自"ins-tea-nt"的缩写形式，是该品牌原有名字的一种变形。这一项目的设计目标是将水果引入到核心品牌之中，强调水果是茶饮的完美伴侣。这一纸杯的功能犹如一个即食方便面纸碗：消费者可以选择先品尝新鲜的水果，随后在杯子里加入热水对茶叶进行冲调。与此同时，设计师还精心地在杯中准备了一个叉/勺，以方便消费者品尝水果。

纸杯咖啡吧

设计机构：新奇设计工作室
完成时间：2011年
设计师：莱斯利·莫里斯
客户：纸杯咖啡吧
国家：澳大利亚

纸杯咖啡吧是一个新落成的意式咖啡吧，坐落于悉尼的斯坦默尔市酒窖的一个古老储藏室内，对于识别的设计方案，设计师运用了专门定制的字母元素，以唤起人们对复古式酒精饮料字体排印工艺的思考。

Paper Cup Coffee

Design agency: Novel
Production date: 2011
Designer: Lesley Morris
Client: Paper Cup Coffee
Nationality: Australian

Paper Cup is a new espresso bar in Stanmore, Sydney. Located in the Stanmore Cellar's old storage room, the identity utilises custom-designed lettering, reflecting retro alcohol typography.

Breyer's Ice Cream

Production date: 2011
Designer: Jinah Lee
Photography: Stephen Han
Client: Breyer's Ice Cream
Nationality: Canadian

布雷耶的冰淇淋店

完成时间：2011年
设计师：真纳·李
摄影师：斯蒂芬·汉
客户：布雷耶的冰淇淋
国家：加拿大

The brand identity itself is much more than just a logo. It should identify the product and the brand, and promote the confidence in the brand. This project re-designed a pre-existing design of Breyer's ice cream package into one that is revolutionary. The new packaging revolutionised the brand identity by creating an effective, unique and fresh new look, staying away from generic design of an ice cream packaging.

这一品牌标识并不仅仅是一个标志。它将成为产品和品牌的象征，有力地提升品牌的公信力。该项目的设计目的是对"布雷耶的冰淇淋"品牌原有的包装设计方案进行重新设计，使其造型获得革命性的转变。全新的包装设计方案通过一个高效、独特、新颖的崭新形象对品牌的识别系统进行了革新，并突破了冰淇淋包装的传统模式。

Zuchi Bistro Café

Design agency: GD86
Production date: 2011
Designer: Keith Morrell
Client: Zuchi Bistro Café
Nationality: British

Asked to rebrand a bistro café, the designer came up with a new brand and applied it to a range of different print items and media. The cup design comes in two different variations with the logo applied in a circular shape to make the brand stand out.

红土咖啡屋餐厅

设计机构：GD86设计工作室
完成时间：2011年
设计师：基思·莫雷尔
客户：红土咖啡屋餐厅
国家：英国

GD86设计工作室应邀为红土咖啡屋餐厅提供品牌重塑方案，设计师们在拟定新品牌方案之后，将其应用到各种印刷品和媒介之中。两种风格相近，外观不同的纸杯设计巧妙地利用圆形形态对标识加以利用，从而有力地突出这一品牌形象。

Ice Queen

Design agency: Iowa State University
Production date: 2011
Designer: Kailee Rose
Client: Ice Queen
Nationality: American

Ice Queen is a package design for ice cream, targeting the demographic of teenage girls. The design reflects the idea that teenage girls walk around with this attitude "it's all about me". The radiating stripes draw the viewer's attention to the centre, along with the tiara illustration and the bright colours demonstrate this concept.

"冰上皇后"

设计机构：美国爱荷华州立大学
完成时间：2011年
设计师：凯丽·罗斯
客户：冰上皇后
国家：美国

"冰上皇后"是一种以少女为消费对象的冰淇淋品牌的包装设计方案。该设计巧妙地体现了少女们"这全部都是我的"的想法。辐射状条纹吸引观者对中心的关注，而细腻的皇冠插图和明亮的色调则更加清晰地阐释了这一设计理念。

Jojo Gelateria

Design agency: Luko Designs®
Production date: 2010
Designer: Lucy Kozozian
Photography: Eric Grigorian
Client: Jojo Gelateria
Nationality: American

Luko Designs® was approached by Mike & Emma Stevens who decided to open their very own Italian ice cream shop in Waukesha, Wisconsin. Their identity and branding pieces came together as a result of their love for bold yet simple graphics that accentuated their yummy gelato treats. Luko Designs® extended the simplicity of the branding to their in-house as well as take-out containers which married the modern to the old-fashioned Americana vintage 50's look.

乔乔冰淇淋店

设计机构：Luko设计工作室
完成时间：2010年
设计师：露西·科佐兹恩
摄影师：埃里克·格里格里恩
客户：乔乔冰淇淋店
国家：美国

迈克与埃玛·史蒂文斯决定在威斯康星州瓦克夏附近开办一家自营意大利冰淇淋店，他们委托Luko设计工作室为其提供设计服务。该项目包括品牌识别与有关品牌部分的塑造，设计师巧妙运用了大胆而简约的图形，以突出该店所经营的美味冰淇淋品牌。Luko设计工作室将该品牌的简约理念延续之内部空间以及外卖容器的设计之中，巧妙地将时尚气息与20世纪50年代的复古风格完美契合。

Little Bird Coffee

小鸟咖啡杯

Production date: 2010
Designer: Madeleine Ward,
Nicole Powell
Photography: Madeleine Ward
Client: Little Bird Coffee
Nationality: American

完成时间：2010年
设计师：马德琳·瓦尔德，
妮可·鲍威尔
摄影师：马德琳·瓦尔德
客户：小鸟咖啡
国家：美国

This logo was developed for a start-up coffee cart located in Boise Idaho. The mission of the business is to create community in the streets while providing quality organic coffee. It was designed to be striking but also incorporate an organic feeling.

这款标志是专门为爱达荷州博伊西新开业的咖啡推车而设计的。咖啡推车事业的使命感在于提供优质的有机咖啡的同时促进街区间更好的沟通。这款设计即独特引人注目又融合了品牌有机的感觉。

Cocínalo Lindo

Production date: 2010
Designer: Maite Cantó
Photography: Maite Cantó
Client: Margie Rullan
Nationality: Puerto Rico

Cocinalo Lindo was an idea generated by a mother of four and one of those daughters. Homemade hearty meals and the option of handcrafted pastries are brought to the schoolyard at the end of the school day. This gives kids with extracurricular activities and their parents in waiting (as well as any student and their parents who want a bite before rushing home) a healthy, homemade option instead of the fast-food route. An identity system, packaging and seasonal menus were designed for Cocinalo Lindo.

"可爱的烹饪"纸杯

完成时间：2010年
设计师：迈特·坎图
摄影师：迈特·坎图
客户：玛吉·卢兰
国家：波多黎各

　　"可爱的烹饪"品牌由一个四个孩子的母亲和其中的一名女儿倾心经营。丰盛的家常菜和纯手工制作糕点在学校即将放学之时被送到校园之中。这一用心打造的品牌为课余时间的孩子们、等候孩子的家长以及他们在飞奔回家之前提供一个健康、家常的饮食服务，从而取代那些不健康的快餐食品。设计师迈特·坎图为该品牌所提供的服务包括识别系统、包装和季节性菜单的设计。

Canopy Coffee

Production date: 2011
Designer: Kelsey McNabb
Photography: Kelsey McNabb
Client: Personal work
Nationality: American

Canopy Coffee was a school project where the designer had to create a brand for a fictional organic coffee company. Through extensive research the designer developed the name, identity, and the design for the coffee cup sleeve and coasters.

树冠咖啡屋

完成时间：2011年
设计师：凯尔西·麦克纳布
摄影师：凯尔西·麦克纳布
客户：个人项目
国家：美国

"树冠咖啡屋"是设计师凯尔西·麦克纳布的一个学校项目，意在为一个虚构的有机咖啡公司创建一个品牌。通过大量的研究，设计师凯尔西·麦克纳布最终开发出该公司的品牌名称、识别系统以及咖啡杯的杯套和杯垫设计方案。

Cassius Bakery Cups

Production date: 2010
Designer: Avigail Bahat
Nationality: Israeli

Cup design for Cassius Bakery and Deli shop. The designer took the simple logo and bright colour to present the style of the shop.

卡西乌斯面包店纸杯设计

完成时间：2010年
设计师：艾维盖尔•巴哈特
国家：以色列

这是为卡西乌斯面包店和熟食商店设计的纸杯。选取了简洁的商标与明亮的颜色来展示商店的风格。

No. 5 Café

Design agency: Novel
Production date: 2007
Designer: Jacqui Lau
Client: No. 5 Café
Nationality: Australian

A vintage stamp was designed to capture the eclectic character of this laneway café in Centre Place, Melbourne. Number 5's identity is now synonymous with Melbourne laneway, café culture and has withstood the test of time. One customer went and got the logo tattooed on his calf.

五号咖啡店

设计机构：新奇设计工作室
完成时间：2007年
设计师：杰奎·刘
客户：五号咖啡店
国家：澳大利亚

五号咖啡店坐落在墨尔本中心广场的一个巷道之上，设计师巧妙地运用一个复古的邮票图案，恰如其分地体现了该咖啡店的折中主义风格。与此同时，五号咖啡店的识别系统与墨尔本巷道、咖啡店文化及其历久弥新的特质相得益彰。曾经一位来到店里的顾客还将该店的标识纹到自己的小腿上。

Magnolia   白玉兰烘焙坊

Production date: 2009
Designer: Tanja Doepke
Client: Personal work
Nationality: German

Magnolia Bakery was the designer's final project to achieve the bachelor degree in graphic and web design. The project should include print, web and packaging. The main goal was to recreate the graphic concept of an existing brand. This means, creation of a new brand identity: new logo, new stationary such as business cards, letterhead and brochures, posters, packaging and a new website.

完成时间：2009年
设计师：唐娅·多普克
客户：个人作品
国家：德国

白玉兰烘焙坊是设计师唐娅·多普克在平面与网页设计专业攻读学士学位的毕业设计作品。该项目包括印刷纸品、网站和包装的设计。其主要目标是对一个现有品牌的平面概念进行重新设计。这就意味着设计师需要为该品牌提供新标识、新文具(如名片、信纸、宣传册、海报等)、包装以及网站的设计方案。

Kabuto Noodles

Design Agency: B&B studio
Production date: 2010
Designer: George Hartley
Client: Kabuto
Nationality: British

Kabuto Noodles heralds a new era for ambient noodles. They have been created to bring restaurant quality noodles to the ambient grocery aisle, to challenge synthetic, poor quality instant snacks. Kabuto is the name of the Samurai warrior's helmet that they also used to eat their food out of prior to battle. The bold brand mark heroes the shape of the helmet and a noodle bowl is "hidden" within the negative space.
The packaging is full of little discoveries to bring the brand to life further and each pack has its own noodle quote, eg. 'when the character of a man is not clear to you, look at his noodles."

卡布托面馆

设计机构：B & B设计工作室
完成时间：2010年
设计师：乔治·哈特利
客户：卡布托面馆
国家：英国

"He Who Knows Others Is Wise. He Who Knows Noodles Is Enlightened."

Step 1) Remove lid and add contents of sachet to pot (using sword to open sachet is not recommended)

Step 2) Add boiling water to top of sleeve, replace lid loosely and wait for 3-4 minutes (opportunity to meditate or practise your karate)

Step 3) Stir, leave for 1 minute, then enjoy straight from the pot or poured into a bowl (if no bowl available, try upside down helmet)

TRUE WARRIORS DON'T LITTER

卡布托面馆的成立预示着面馆环境进入了一个崭新的时代。该面馆力图为附近的杂货店主提供餐厅品质的面条服务，从而取代那些质量较差的综合型快餐店。卡布托原指武士的头盔，在古代武士们经常在战斗之前用它来盛放食物。设计师大胆地将武士的形象与头盔融为一体，并在负空间中"隐藏"一个面碗图案，可谓匠心独运。
包装的设计精致而巧妙，其中所设置的小发现，轻松地拉近了品牌与生活之间的距离，每个包装上均设有其独特的面条引言，例如，"如果您还尚不清楚一个人的性格，那么请观察他选的面条吧。"

Ai Garden

Production date: 2010
Designer: PirayaRuangpungtong
Photography: Korkiat Kittisoponpong, Silasak Kaewmaneerat
Client: Ai Garden
Nationality: Thai

Ai Garden (garden of love) is a bakery shop with relaxing atmosphere, tasty bakery, ice cream and beverage, located in Nihonmachi, the lifestyle mall in Bangkok. The corporate identity design contains all promotional design such as logo, mascot, menu, pop-up menu, vinyl flag, signage, business card, paper cup and tissue paper.

爱情花园

完成时间：2010年
设计师：皮拉娅–阮彭东
摄影：科琪亚·吉蒂思颇彭，斯拉萨·克凯渥曼尼拉特
客户：爱情花园
国家：泰国

爱情花园是一家烘焙店，坐落在曼谷一个名为日本町的商场内，拥有轻松、休闲的内部环境，主要出售美味的烘焙类食品、冰淇淋、饮料等。这一企业标识设计方案包括所有的宣传设计，诸如标志、吉祥物、菜单、弹出式菜单、乙烯标志、标牌、名片、纸杯和纸巾等。

Smooth Coffee

Production date: 2009
Designer: Andrei Antonescu
Client: Smooth Coffee
Nationality: Romanian

Identity for a shop/boutique with a large array of services, ranging from fruit "smoothies" to light food and coffee. The logo was meant to be fresh and the packaging is simple.

醇和咖啡店

完成时间：2009年
设计师：安德烈・安东内斯库
客户：醇和咖啡店
国家：罗马尼亚

该项目是设计师安德烈・安东内斯库专为一个店铺/精品店而设计的识别系统，这家精品店内为顾客提供多样化的餐点服务，涉及水果冰沙、清淡食品和咖啡等。其中，标志的设计力图突出新鲜这一主题，包装风格简约、清新。

Cafexpress

极速咖啡厅

Production date: 2011
Designer: Jasmine Holm
Client: Cafexpress
Nationality: Australian

完成时间：2011年
设计师：杰斯敏·霍尔姆
客户：极速咖啡厅
国家：澳大利亚

Cafexpress is a mobile coffee company based in Sydney, Australia that provides quality coffee that keeps you moving. The purpose of the design is to promote on-the-go, quality coffee in a warm and welcoming manner. The arrow symbol represents the speed of service, the stock choice represents the quality and the bunting support graphics represent a guaranteed fun, enjoyable visit to Cafexpress every time!

极速咖啡厅是一个移动式咖啡公司，坐落在澳大利亚的悉尼，其服务宗旨是为行驶的客户提供优质的咖啡。该设计的目的是以一种温馨、热情的方式突出便携式优质咖啡的特色。箭头符号象征着服务的速度，箭杆象征着品质，而作为辅助图案的旗布则象征着趣味性，即为每次来到极速咖啡厅的顾客带来美好心情。

אֲרוֹמָה

אספרסו בר

Aroma

Production date: 2009
Designer: Yotam Bezalel
Client: Aroma
Nationality: Israeli

The designers found the solution by forming a new media strategy that underscores the endless range of goods which Aroma has to offer. The graphic language included various illustrations on the simple line of products which depicted: music, health, involvement in community service, recycling, etc. The project included redesigning the packaging, menus, advertising and marketing materials.

香氛品牌店

完成时间：2009年
设计师：犹塔姆·比扎莱尔
客户：香氛品牌店
国家：以色列

设计师犹塔姆·比扎莱尔等人构思了一个全新的媒体战略以突出香氛品牌店所经营的丰富商品。该品牌的平面语言包括各种以产品的简单线条为基础的插画，以描述音乐、健康、参与社区服务、回收等为主题。这一项目包括包装、菜单、广告和营销材料的重新设计。

The Lebanese Ice Cream Since 1936
البوظة اللبنانية منذ ١٩٣٦

Bachir Icecream

Design agency: Design Mesh
Production date: 2010
Designer: Elie Abou Jamra
Client: Bachir Icecream in Beirut
Nationality: Lebanese

Bachir has been around for decades, and this is the trigger of a purely typographic logo. Retaining the brand's authentic red colour, the logotype reflects the urban essence of Bachir through the crooked and irregular forms of the type.

贝希尔冰淇淋店

设计机构：设计网络工作室
完成时间：2010年
设计师：埃利·阿布·加马拉
客户：贝鲁特的贝希尔冰淇淋店
国家：黎巴嫩

贝希尔冰淇淋店已经拥有几十年的创办历史，其委托设计网络工作室为其设计一个简洁的印刷字体标识。设计师埃利·阿布·加马拉保留了该品牌原有的鲜红色调，利用弯曲而不规则的字体形态构成标识，充分地彰显出贝希尔冰淇淋店的都市本质。

Coffee & Kitchen

Design agency: moodley brand identity
Production date: 2011
Creative director: Mike Fuisz
Designer: Nicole Lugitsch
Photography: Marion Luttenberger
Client: Coffee & Kitchen
Nationality: Austrian

Situated in a business district of Austria's second biggest city Graz, the new restaurant Coffee & Kitchen restaurant brings culinary pleasures to the daily office life. The colour world in black and white (logo) combined with brown (wrapping papers, menu cards, food packaging...) determines the interior design as well as the corporate design. moodley brand identity has consciously avoided the branding printed material; however there are different logo stickers that convey the image of a relaxed and informal restaurant atmosphere (even the signboard symbolises a sticker). The used handwriting font intensifies this feeling even more.

"咖啡与厨房"餐厅

设计机构：莫德利品牌标识设计工作室
完成时间：2011年
创意总监：：迈克·福瑞兹
设计师：尼科尔·卢奇特斯科
摄影师：马里昂·卢腾伯格
客户："咖啡与厨房"餐厅
国家：奥地利

新成立的"咖啡与厨房"餐厅坐落在奥地利第二大城市格拉茨的一个商务区内，以为日常办公一族带来烹饪的乐趣为服务主题。黑色、白色（标识）与棕色（包装纸、菜单卡、食品包装）的完美结合，奠定了室内设计以及企业识别设计的基础。莫德利品牌标识设计工作室有意避开品牌印刷材料的运用，而是巧妙地运用各种标识粘贴以充分地传达出该餐厅的轻松、自在氛围（即便广告牌也象征着一张贴纸）。手写字体更加强化了这种自在之感。

068

Cone Kings Frozen Yogurt

Design agency: In-house
Production date: 2011
Creative director: Arjan van Woensel,
Mathew Metcalfe
Designer: Arjan van Woensel
Client: Cone Kings Ltd
Nationality: New Zealand

This is a stand-out paper cup design for a new premium frozen yogurt. It achieves the feel of a luxury brand by the use of colours and filigree elements, engraving-style illustrations, elegant understated typography, and by using the brand name as a signature only ("by Cone Kings"). Bold use of colour really makes a difference from all competition that mostly uses white and pastel colours.

"圆锥王国"冷冻酸奶包装

设计机构：In-house设计工作室
完成时间：2011年
创意总监：阿尔简·凡·沃恩赛尔，
马修·梅特卡夫
设计师：阿尔简·凡·沃恩赛尔
客户："圆锥王国"有限公司
国家：新西兰

该项目是In-house设计工作室专为一个优质冷冻酸奶而提供的独立式纸杯设计方案。设计师巧妙运用各种色彩、金银丝细工元素、雕刻风格的插图、高贵而低调的字体以及有关品牌名称的签名（由"圆锥王国"有限公司提供），完美塑造了一个奢华、高贵的食品品牌。大胆的色彩运用使其与市场上惯用白色和粉彩的同类商品鲜明地区分开来，真正实现独树一帜的效果。

Mathew Metcalfe
MASTER BLENDER

MATHEW@CONEKINGS.COM PHONE 021 684 496

The CONE KINGS

WWW.CONEKINGS.COM 188 JERVOIS RD AUCKLAND

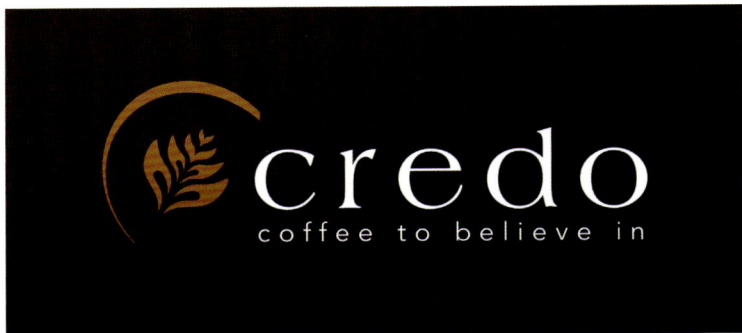

Credo Coffee

Design agency: WeAreAllConnect
Production date: 2009
Designer: Jesse Campbell & Curtis Sorensen
Client: Credo Coffee
Nationality: Canadian

The concept for Credo Coffee was simple: a café that focuses on quality from the grower to the cup. The result was a rosetta icon, reflecting the art that happens in every cup.

克雷多咖啡

设计机构：WeAreAllConnect设计工作室
完成时间：2009年
设计师：杰西·坎贝尔，柯蒂斯·索伦森
客户：克雷多咖啡厅
国家：加拿大

克雷多咖啡厅的设计理念非常简约，重点要突出该咖啡厅对咖啡豆从种植到最终成为杯中物这一过程的品质关注。设计师最终设计了一个罗塞塔图标，以彰显出每个杯子中所展现的艺术之美。

Fattigmann

Design agency: Sukker Design
Production date: 2009
Designer: Ellen Katrine Kristensen, Alexander Qual
Photography: Sukker Design
Client: Umoe Restaurant Group AS
Nationality: Norwegian

This is a complete repositioning and redesign of the bakery chain. The designers developed the brand experience from conceptual development, identity, packaging and interior design.

法蒂格曼烘焙坊

设计机构：苏克尔设计工作室
完成时间：2009年
设计师：埃伦·卡特琳·克里斯腾森，亚历山大·库尔
摄影师：苏克尔设计工作室
客户：Umoe餐厅集团
国家：挪威

该项目是苏克尔设计工作室专为一个连锁烘焙坊而提供的品牌重新定位与设计方案。设计师埃伦·卡特琳·克里斯腾森与亚历山大·库尔倾力打造的这一品牌开发方案包括概念的开发、识别系统、包装以及内部空间的设计。

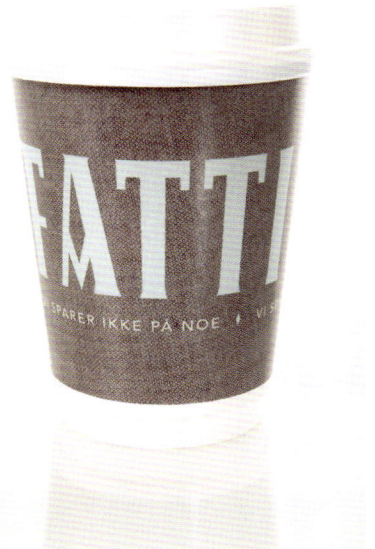

Froberry Frozen Yogurt

Design agency: WeAreAllConnect
Production date: 2011
Designer: David Lawlor
Client: Froberry
Nationality: Irish

This is a paper cup design for Froberry. It is great tasting frozen yogurt. It's made with pure natural yogurt and the freshest ingredients from great Irish farms. Froberry is tangy and gluten free – at sub 100 calories a portion; it's so much lighter than ice cream!

弗洛波里冷冻酸奶

设计机构：WeAreAllConnect设计工作室
完成时间：2011年
设计师：大卫·劳勒
客户：弗洛波里冷冻酸奶
国家：爱尔兰

该项目是设计师大卫·劳勒专为弗洛波里冷冻酸奶而提供的纸杯设计。弗洛波里冷冻酸奶是一款风味独特的饮品，其以纯天然的酸奶和大型爱尔兰农场的最新鲜食材为原料。弗洛波里冷冻酸奶味道香浓，不含麸质，远低于100卡路里，其热量要远远低于冰淇淋。

I made it

Production date: 2010
Designer: Sveta Fedarava
Client: Yutopa. All Natural Frozen Yogurt
Nationality: Canadian

The main goal was to design a paper cup that would be fun and reflect light and cheerful personality of the brand. The designers also wanted to incorporate slogan of the company: "i made it", which works quite well on the back of the paper cup.

"我成功了"

完成时间：2010年
设计师：诗微达·费达拉瓦
客户："Yutopa"全天然的冷冻酸奶
国家：加拿大

该项目的主要目标是设计一个纸杯，在确保其趣味性的同时彰显出该品牌轻盈、欢快的个性。与此同时，设计师诗微达·费达拉瓦等人还巧妙地将该公司的宣传口号"我成功了"添加到纸杯的后身，真正做到了匠心独运。

i made it™

yutopia.ca

yutopia®
all natural frozen yogurt

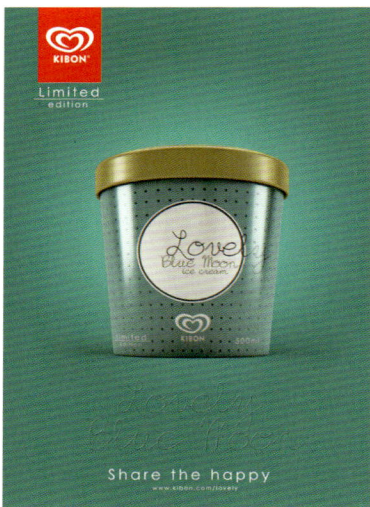

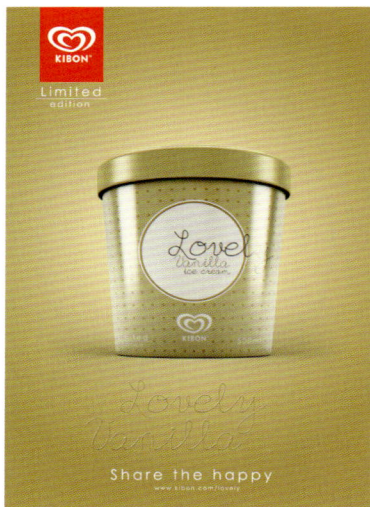

Lovely Ice Cream

Design agency: WPIXY STUDIO
Production date: 2011
Designer: Felipe Henrique dos Santos
Client: Kibon
Nationality: Brazilian

The idea was to create a new line of products limited, inspired by a traditional cup design for the ice cream, but with a modern art, simple, vibrant colours. It was not tiring to the eyes of consumers in this way calling attention.

可爱的冰淇淋

设计机构：WPIXY设计工作室
完成时间：2011年
设计师：菲利普·恩里克·多斯·桑托斯
客户：Kibon冰淇淋制造商
国家：巴西

该项目的设计理念是创造一个全新的限量版系列产品，其设计的灵感源自冰淇淋的传统纸杯设计风格，设计师巧妙运用时尚艺术处理手法，搭配简约、活泼的色调，成功吸引消费者的注意力，并毫无视觉疲劳之感。

marronerosso

Marrone Rosso

马诺尼·罗索咖啡

Production date: 2011
Designer: Yotam Bezalel
Client: Marrone Rosso
Nationality: Israeli

完成时间：2011年
设计师：约塔姆·比扎莱尔
客户：马诺尼·罗索咖啡
国家：以色列

Marrone Rosso is a prestigious branch of Aroma Israel located in Europe. The purpose was to brand a new coffee chain, an extension of Aroma Israel in Europe.
On one hand, the designer wanted to preserve a bit of Aroma's image, while on the other hand creating a brand which is "high-class" and prestigious.
The project included: design of logo which was taken from the coffee bean shape and colour, packaging, ads, wallpaper, menus and flyers.

马诺尼·罗索咖啡是欧洲"香氛以色列"咖啡的一个著名子品牌。该项目的目的是为这一全新的咖啡连锁店进行品牌设计。
一方面，设计师希望能够保留一些"香氛以色列"咖啡品牌的形象，另一方面力图打造一个"高级"、享有盛名的品牌。
该项目包括：标识的设计（取材自咖啡豆的形状与色调）、包装、广告、墙纸、菜单和宣传页。

Old Man's Dream

Production date: 2009
Designer: Avasiloiei liviu
Client: Personal work
Nationality: Romanian

It's nice and refreshing to drink your morning coffee on the way. And to the perfection of the moment contributes not only a good coffee, but also a friendly and comfortable design of the take-away cup. That's why the designer created Old Man's Dream, a personal "sensory" identity for the take-away cup.

老人的梦想

完成时间：2009年
设计师：埃瓦斯洛·利维乌
客户：个人作品
国家：罗马尼亚

在早晨上班的路上品尝一杯浓香的咖啡的确是一件很惬意、清爽的事情。而这种惬意单凭一杯香浓的咖啡并不能够实现，一个友好、舒适的咖啡杯设计所扮演的重要角色也绝不容小觑。这也是设计师埃瓦斯洛·利维乌创作这个项目——"老人的梦想"的目的所在，精致的设计风格完美地传达出其对便携式纸杯的个性化感知与理解。

Oriental Bakery

Production date: 2008
Designer: Mohammed Al-Mousa
Client: Oriental Bakery
Nationality: Jordanian

The Oriental Bakery goes in line with the colours of soft, white bread. That is why the main colours of the design consist of the light beige of wheat and the dark brown colour of crust to give the perfect contrast of baking. The golden wheat in the logo is slightly curved to give a smooth and delicate shape. The whole packaging & stationary theme give a simple and smooth appearance which is exactly what Oriental Bakery is all about.

东方烘焙坊

完成时间：2008年
设计师：穆罕默德·阿尔－穆萨
客户：东方烘焙坊
国家：约旦

东方烘焙坊以经营色彩柔和的白色面包为主要特色。这种色调同时也奠定了设计师穆罕默德·阿尔－穆萨所提供的设计方案的基础，设计师巧妙运用小麦的浅米色和面包皮的深棕色，使之与烘焙过程形成完美的对比。标识中的金色小麦经过轻微的弯曲之后呈现出一个圆润、精致的轮廓。整个包装和文具的设计主题强调简约和流畅的格调，而这恰恰也与东方烘焙坊所提倡的理念相得益彰。

Picnic

Design agency: Sukker Design
Production date: 2008
Designer: Alexander Qual, Reidar Oksavik
Client: SSP Select Service Partner
Nationality: Norwegian

For this airport café chain the designers created the full brand experience from conceptual development, identity, packaging to interior design.

野餐

设计机构：苏克尔设计工作室
完成时间：2008年
设计师：亚历山大·库尔，雷达尔·奥克撒韦克
客户：精选餐饮管理有限公司
国家：挪威

该项目是苏克尔设计工作室专为一个机场连锁咖啡厅而设计的品牌方案，这一全方位的品牌塑造方案包括品牌概念的开发、识别系统的设计、包装和室内设计。

Decaf
Shots
Syrup
Milk
Drink

a touch of earth NATURALLY CONSCIOUS CAFE

A Touch of Earth: To-Go Coffee Cups

Production date: 2011
Designer: Justin Marimon
Client: A Touch of Earth
Nationality: American

A Touch of Earth is a small coffee shop located in downtown Columbus, Ohio in the United States. The company's main focus is being an environmentally friendly place to get coffee and other café items in a convenient downtown setting. This cup was designed as a simple and iconic representation of the brand. Being that coffee cups are consumed daily by many people, these cups are created using 100% post consumer materials and are, in turn, 100% biodegradable. They do not harm the environment in any way. The coffee sleeve would also be biodegradable and be a direct reversal of the colours found on the cup beneath.

轻触大地：便携式咖啡杯

完成时间：2011年
设计师：贾斯汀·马里蒙
客户：轻触大地咖啡屋
国家：美国

"轻触大地"是一个小型咖啡屋，坐落在美国俄亥俄州哥伦布市中心。该公司的主要经营理念是以一个便捷的市中心位置为背景，在一个环保的环境中为顾客提供咖啡以及其他咖啡馆产品。这一纸杯的设计风格崇尚简约而醒目，是该品牌的有力象征。考虑到每天人们要消耗大量的咖啡杯，因此，这些咖啡杯以消费者使用后的材料为原料，并完全可以进行生物降解。如此，这种设计不会对环境造成任何的伤害。除此之外，咖啡杯的杯套也同样可进行生物降解，其色彩与杯底的颜色形成鲜明的对比。

90 Degrees Café

90度咖啡馆

Design agency: Leaf Design Pvt. Ltd.
Production date: 2009
Client: 90 Degrees Café
Nationality: Dubai

设计机构：叶片设计有限公司
完成时间：2009年
客户：90度咖啡馆
国家：迪拜

Ninety Degrees, a coffee shop in Dubai, is in the heart of Jumeriah Beach Residence, a unique café shop, by giving the customer an experience of coffee drinking in a fun, relaxing yet information manner. The strategic location and a unique name created a lot of avenues to explore. The challenge was to develop a new identity which should differentiate the café with its funky, classy design with touch of quirkiness and an ambience to relax. A Tangram Twister was used as it is very dynamic and quick, entirely relatable to the fun, quirkiness and intelligence that the café wanted to associate with. It was created as a logo and identity to give it a very casual fun feel.

90度咖啡馆坐落在迪拜，位于卓美亚海滩别墅的中心，它是一个风格独具的咖啡店，凭借一种有趣、轻松、信息化的方式为顾客营造出前所未有的感官体验。优越的地理位置和独特的名称为其提供了多种开发途径。设计的挑战是如何开发一个全新的识别系统，使之凭借独特而经典的设计风格以及千变万化、轻松休闲的环境与其他同类咖啡店鲜明地区分开来。设计师巧妙运用七巧板的形象，利用其充满动感与敏捷的特质象征着该咖啡店妙趣横生、富于变化、智能灵活的独有魅力。此外，这一七巧板作为标识和识别设计的主要元素，自然地流露出轻松、休闲之感。

Feminist Coffee

Design agency: Anders Wallner
Production date: 2010
Designer: Anders Wallner
Client: Feministiskt Initiativ
Nationality: Swedish

These take-out-coffee cups are made by Anders Wallner after his own idea and art direction for Swedish feminist party, "The Feminist Initiative" ("F!"), and was made to reach out to stressed and busy big city people who would never take time to grab or read a flyer from a party or anything for that matter, but by giving them free coffee you'd not only get them to take the info, they'd read it and maybe most important they'd walk around displaying both the logo and info. And really who'd say no to free coffee?

The cups feature the three most important issues for the party, such as Equal salaries, Shorter workdays and "A new greener way" and therefore present the party in a "bigger picture" kind of way new electors would not see them in otherwise.

女权主义者咖啡杯

设计机构：安德斯·瓦尔纳设计工作室
完成时间：2010年
设计师：安德斯·瓦尔纳
客户：女性主义行动先锋
国家：瑞典

这些便携式纸杯是设计师安德斯·瓦尔纳专为瑞典女性主义政党"女性主义行动先锋"而倾力打造，并在整个设计过程中担任艺术指导。同时，该系列纸杯还以疲于奔命的都市人群为目标，因为这些人群平时根本没有时间接收或阅读有关某个党派或团体的宣传单，然而，一杯免费的咖啡不仅能够有效传达信息，更重要的是它还可能将这一标识和信息四处散播。况且，谁又会拒绝免费的咖啡呢？

这一系列咖啡以"女性主义行动先锋"政党关注的三个重大问题为主题，例如"薪资平等"、"缩短工作时间"和"一个更为环保的新方式"，以一种前所未有的"大局"方式展现这一政党的本质。

*iceshake bar

JAY JAY'S

JayJay's Iceshake Bar

Design agency: urbe
Production date: 2011
Designer: Piotr Juszczak
Client: JayJay's Iceshake Bar
Nationality: Polish

The aim was to distinguish JayJay's from other coffee bars. The designers foccused on the wide range of iceshakes instead of coffee symbols. They created the logotype that drinks itself and is an innocent promise of an iceshake delight.
JayJay's quasi face is emotional – it shows thirst fulfilment. The designers worekd out the scheme for shoving people having a sip of their favourite iceshake – a moment with JayJay's.

杰杰的冰摇吧

设计机构：urbe设计工作室
完成时间：2011年
设计师：彼得·杰斯兹克扎克
客户：杰杰的冰摇吧
国家：波兰

该项目的设计目的是将杰杰的冰摇吧与其他咖啡吧进行鲜明地区分。来自urbe设计工作室的设计师们力图突出该品牌丰富的产品种类而非咖啡标志。设计师们倾力打造的标识以开怀畅饮的形象呈现，匠心独运而又不失可爱之风。杰杰的冰摇吧的标识形象充满强烈的感情色彩，展现了对饮品的渴望。设计师最终打造的这一方案意在吸引顾客纷至沓来，争相在杰杰的冰摇吧中畅饮自己喜欢的冰摇系列饮品。

*iceshake bar

JAY JAY'S

contemporary lifestyle

Tazza Doro

Production date: 2009
Designer: Yotam Bezalel
Client: Tazza Doro
Nationality: Israeli

Tazza doro has no branches or specific branding. Nonetheless, the designers wanted to create some kind of diverse graphic language to fit the high-standard of the place. The project included a campaign of advertisements and posters promoting the "cinema Tazza" festival, menus, packaging, advertisements, etc.

塔扎·多龙

完成时间：2011年
设计师：尤太·贝扎勒
客户：塔扎·多龙
国家：以色列

塔扎·多龙既不是个十分有名气的品牌，也没有其他分店。然而设计师想运用各种各样的平面设计语言来展现塔扎·多龙高水准的服务。这个设计项目包括推广"塔扎电影院"节日活动的一系列广告和海报、菜单、打包袋、和广告宣传册等相关产品。

CrimsonCup

Design agency: Salvato Coe + Gabor
Production date: 2006
Designer: Mark Gormley
Client: CrimsonCup Coffee & Tea
Nationality: American

CrimsonCup Coffee & Tea has roasted and packaged quality, great-tasting coffee in Columbus, Ohio for 20 years and thrives by teaching independent business owners how to successfully run their own coffee shops.

深红咖啡杯设计

设计机构：撒瓦图·蔻伊与加博设
计工作室
完成时间：2006年
设计师：马克·戈姆利
客户：深红杯咖啡与茶品牌
国家：美国

深红杯咖啡与茶拥有烤制与包装的优
良品质，来自美国俄亥俄州哥伦布市
的上等美味咖啡，已经有20年的历
史，通过指导独立的业主如何成功的
经营其咖啡店而得到蓬勃发展。

Square Paper Cup

Design agency: O-D Studio
Production date: 2010
Designer: Daniel Powter
Client: Square Coffee Bar
Nationality: British

"Square" is the name of a coffee brand. Orange and black are the predominant coulours of this brand of coffee shop. Broad coulour scheme forms a strong vision impact. To follow the design style of the shop, the designer Daniel Porter adopts orange as the main coulour of the paper cup combined with black as a perfect combination. This coulour design together with the shop design has a great appeal. Customers can bring the paper cups everywhere and thus they become a free medium of advertisement.

广场纸杯

设计机构：O-D设计工作室
完成时间：2010年
设计师：丹尼尔·波特
客户：广场咖啡酒吧
国家：英国

每当提起Square，人们就会联想到咖啡。店面的设计主要采用橘色和黑色。这样明丽的色彩设计给人以强烈的视觉冲击感。为了延续店面的设计特点，平面设计师在纸杯设计时，在橘色底色上采用黑色圆圈图案以配合黑色商标的设计。这样的设计会让人们过目不忘。顾客可以将纸杯带到城市的各处，所以纸杯可谓是"长了双脚"的广告。

Red Espresso Bar

Design agency: Anna Geslev Brand Design
Production date: 2011
Designer: Teddy Cohen, Nir Tober, Anna Geslev
Client: Red Espresso Bar LTD
Nationality: Israeli

Red Espresso Bar, an up-and-coming international café brand, represents a new trend — specialty-coffee shop. The designers focused on creating a long-lasting, modern, flexible, bold and "red" visual language that would apply to every media. Red's photographic style derived from the gadget world to put the focus on quality, desirable, and yet accessible products.

红色意式咖啡吧

设计机构：安娜·杰斯雷夫品牌设计工作室
完成时间：2011年
设计师：泰迪·科恩，尼尔·托博，
　　　　安娜·杰斯雷夫
客户：红色意式咖啡吧有限公司
国家：以色列

红色意式咖啡吧是一个极具发展潜质的国际咖啡厅品牌，代表特色咖啡店的全新发展潮流。安娜·杰斯雷夫品牌设计工作室力图打造一个历久弥新、时尚、灵活、前卫的"红色"视觉语言，并将其广泛地应用到各种媒介之中。纸杯的摄影内容取材自各种小器具，重点强调高品质、受人欢迎同时容易获得的商品。

Riva Café Cup

Production date: 2008
Designer: Kevin Leung
Client: Riva Café
Nationality: Australian

This is a rebranding of a back alley café located in the City of Melbourne. The goal is to create simple cups which can catch the attention of people.

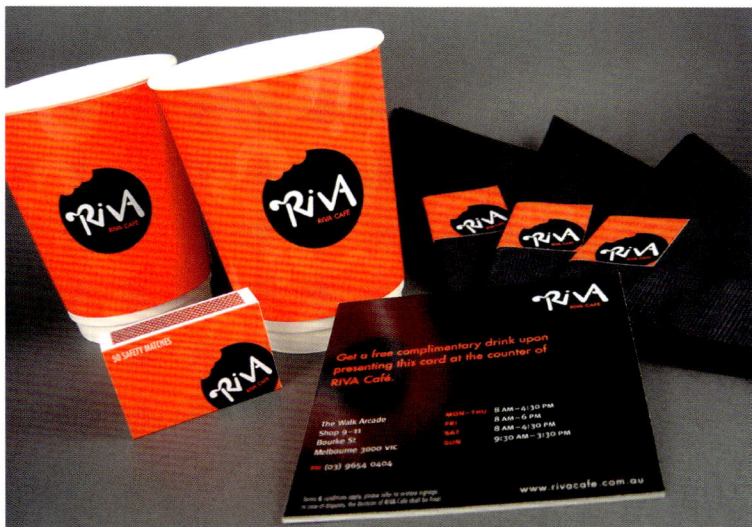

里瓦咖啡杯

完成时间：2008年
设计师：凯文·莱翁
客户：里瓦咖啡
国家：澳大利亚

该项目是设计师凯文·莱翁专为一个坐落在墨尔本市的小巷咖啡屋而设计的品牌重塑方案。设计的目标是创建一个简约而极具视觉感染力的咖啡杯，使之能够轻松捕捉消费者的目光。

One Tree Coffee

Design agency: Boheem Design
Production date: 2011
Designer: Claire Bonnor, Claire Robinson, Chantel De Sylva
Client: One Tree Coffee Co.
Nationality: Australian

One Tree Coffee Co. is a boutique Espresso Bar in Newcastle NSW. They offer a small range of quality coffee blends, including their very own iced coffee – bottled in the store and sold within a day. The overall aesthetic takes reference from the coastal city's coal and steel industries as well as the old railways used for transport of these goods. The design components use elements found in vintage train tickets and railway signs.

一棵树咖啡公司

设计机构：波希姆设计工作室
完成时间：2011年
设计师：克莱尔·波诺尔，克莱尔·罗宾逊，尚特尔·德·萨尔瓦
客户：一棵树咖啡有限公司
国家：澳大利亚

一棵树咖啡有限公司是一个精品意式咖啡吧，坐落于新南威尔士州纽卡斯尔市。该咖啡吧经营小范围的优质混合咖啡，其中包括他们自己加工并在店内出售的瓶装冰咖啡。该纸杯的整体美学理念取材自沿海城市的煤炭和钢铁等行业以及输送这些产品的老式铁路。设计的组成部分包括老式火车票和铁路标志中常见的元素。

The Hayes Café

海耶斯咖啡吧

Design agency: Paco.Branding & Design
Production date: 2011
Designer: Judy Moosmueller
Client: The Hayes café
Nationality: Australian/German

设计机构：柏高品牌塑造与设计工作室
完成时间：2011年
设计师：朱迪·莫斯穆勒
客户：海耶斯咖啡吧
国家：澳大利亚/德国

The identity was developed with a warm, traditional feel to underline the homemade aspect of this samll catering business. Care, honesty and dedication to quality are integral to the business and reflected in the design. The symbol visually links to having a "cherry on top" meaning to have something good. The bright red cherry-shaped sticker functions as a cost-effective device that may be used on items diffcult to print. A collage with illustrated food products was developed for other collateral to compliment and complete the visual indentity.

温暖、传统的识别设计风格强有力地突出了这一小型餐饮企业的家常烹饪技巧。关怀、诚实与优质构成了该公司的经营理念，并在整个设计中完美地流露出来。设计师巧妙地在标志的上方设置樱桃的形象，寓意美好的事物。醒目的鲜红色樱桃形贴纸作为一个性价比较高的图案，能够轻松摆脱难以印刷的烦恼。除此之外，设计师还精心设计了一个绘有食品的拼贴画，使其作为整个视觉识别系统的附属部分。

089

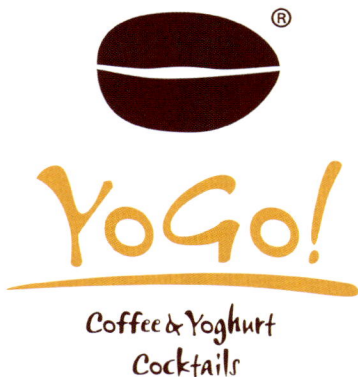

YoGo! Coffee & Yoghurt Cocktails

优格咖啡和酸奶鸡尾酒

Design agency: urbe
Production date: 2011
Designer: Piotr Juszczak
Client: YoGo!
Nationality: Polish

设计机构：urbe设计工作室
完成时间：2011年
设计师：彼得·杰斯兹克扎克
客户：优格公司
国家：波兰

The designers believe that both things have the same power of influence. YoGo! is not only coffee, but also yoghurt cocktails, frozen if you like.
They created name "YoGo!" to make it memorable thanks to hidden words like: You, Yogurt, Go! (fast/takeaway). The designers gave the logotype an extra twist with a beauty spot converted into sign original ® inspired by Marylin Monroe's sensuality.

urbe设计工作室坚信两种事物会有相同的影响力。优格不仅以咖啡闻名，其生产的酸奶鸡尾酒（亦可冷冻）也同样深受消费者的欢迎。
设计师精心打造的"优格"这一名称，源自英文意为"你，酸奶，立即购（快速/外带）"的缩写。
设计师巧妙地将标识进行一定程度的变形，并模仿玛丽莲·梦露性感的美人痣，在标识的上方添加"®"标识，匠心独运。

Casa Coffee

Design agency: Sole Designer
Production date: 2009
Designer: Piotr Juszczak
Client: College Project
Nationality: Australian

Casa Coffee combines coffee from around the world with an easy-to-understand colour coded system. The coffee beans are housed in Jute material that brings an essence of authenticity and coffee experience back to the customer.

卡萨咖啡

设计机构：个人设计
完成时间：2009年
设计师：彼得·杰斯兹克扎克
客户：学院项目
国家：澳大利亚

卡萨咖啡品牌结合了世界各地的咖啡风味，拥有一个易于理解的彩色编码系统。放置在黄麻材料中的咖啡豆将为顾客带来一种纯粹、真挚之感。

mała
czarna

Mala Czarna

Design agency: Zaven
Production date: 2009
Designer: Arthur Krupa
Client: Mala Czarna
Nationality: Polish

The project is a conceptual branding and product packaging for a café. The idea was to communicate the warm and lazy feeling one has while enjoying a freshly brewed coffee. The design features four female characters, each representing a country which is home to coffee beans (Alma — Columbia, Desta — Ethiopia, Rita — Brasil, Shanti — India).

马拉·洽尔纳咖啡厅

设计机构：Zaven设计工作室
完成时间：2009年
设计师：阿瑟·克鲁帕
客户：马拉·洽尔纳咖啡厅
国家：波兰

该项目是Zaven设计工作室专为马拉·洽尔纳咖啡厅所提供的概念品牌塑造和产品包装设计方案。设计的理念是有力地传达出顾客在这一温馨、慵懒的氛围中品尝现煮咖啡的惬意之感。该设计以四位女性角色为主题，其中，每位女性代表一个咖啡豆生产国（即阿尔玛代表哥伦比亚，德士达代表埃塞俄比亚，丽塔代表巴西，尚蒂代表印度）。

Adir Dairy

Design agency: Blend-It Design
Production date: 2009
Designer: Assaf Cohen, Ernesto Bijovsky
Client: Adir Dairy
Nationality: Israeli

Adir Milk is the largest premium goat milk-producing dairy in Israel. The dairy, which is located in the Upper Galilee, has served as an experiment to create a dairy milk experience that reflects a feeling of tradition combined with an ultimate taste experience. The search for a design language that relies on the best of French cheese design traditions has led the designers to a world of primal emotions, simplicity, and elegance, conveying freshness, quality, and locality.

安迪尔乳品

设计机构：Blend-It设计工作室
完成时间：2009年
设计师：阿萨夫·科恩，埃内斯托·比约夫斯基
客户：安迪尔乳品
国家：以色列

安迪尔乳品是以色列最大的优质羊奶制品加工商。这一奶制品加工商坐落在上加利利地区，是打造全新乳制品体验的一次伟大尝试，将传统之感与绝妙的味觉体验完美结合。对于设计语言的探索，设计师以最完美的法式奶酪设计传统为基础，力图引领观者进入一个单纯、简约、优雅的情感世界，促使他们体会清新、优质之感以及浓厚的地域特色。

Full Stop Bistro Branding

Production date: 2010
Designer: Millie Rose Cordingley
Client: Full Stop
Nationality: British

The cup design is part of the branding for a punctuation-themed Bistro called "Full Stop". Every aspect of the branding used punctuation within it; for example the pattern used on the takeaway cup is made from a repeated exclamation point.

句点小酒馆品牌设计

完成时间：2010年
设计师：米莉·罗斯·科丁利
客户：句点小酒馆
国家：英国

这一纸杯设计方案是句点小酒馆品牌塑造的一个部分，句点小酒馆强调以标点符号为主题。该酒馆品牌设计方案的每个部分均运用了酒馆中的标点元素，例如，便携式纸杯上的图案巧妙运用了一个重复的惊叹号。

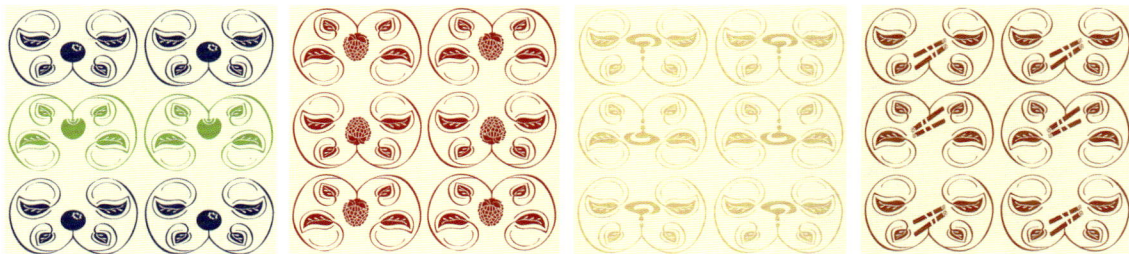

Quaker Oats

Design agency: Staffordshire University
Production date: 2010
Designer: David Salt
Client: D&AD
Nationality: British

This is a design for a new "Chilled Creamy Oats" product, aimed at young professional women looking for a delicious healthy snack. The name "Natural Charm" represents the natural ingredients and elegance of the design. The colour and centrepiece of the repeated pattern changes on each flavour to distinguish the fruit content.

桂格燕麦

设计机构：斯塔福德郡大学
完成时间：2010年
设计师：大卫·萨尔特
客户：英国设计与艺术指导协会
国家：英国

该项目是设计师大卫·萨尔特专为桂格燕麦品牌新推出的一款"冰鲜奶油燕麦"而提供的设计方案。该款燕麦产品以善于挖掘美味健康小吃的年轻职业女性为主要消费对象。"天然的魅力"这一名称象征着纯天然成分和优雅的设计。此外，每种水果口味的包装色调不同，却全部遵循图案重叠的设计理念。

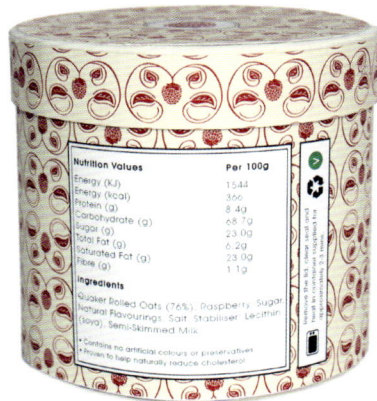

Scoops Ice Cream Shoppe

Production date: 2010
Designer: Katie Lewis
Client: Scoops Ice Cream
Nationality: American

Scoops Ice Cream Shoppe is a proposed re-branding for The Chocolate Shoppe Ice Cream in Madison, Wisconsin. The logo uses custom-designed lettering and imagery that references calligraphy. The colours and patterning reflect the lighthearted atmosphere present in the The Chocolate Shoppe.

速酷冰淇淋专柜

完成时间：2010年
设计师：凯蒂·刘易斯
客户：速酷冰淇淋
国家：美国

速酷冰淇淋专柜是设计师凯蒂·刘易斯专为威斯康星州麦迪逊市巧克力冰淇淋专柜而设计的品牌重塑方案。该品牌的标识运用了量身定制的字母和图像，设计的灵感源自设计师对书法的参考。配色方案和图案设计完美地彰显出巧克力冰淇淋专柜中愉悦的环境氛围。

Ivry Dairies

Design agency: Blend-It Design
Production date: 2008
Designer: Assaf Cohen, Ernesto Bijovsky
Client: Ivry Dairies
Nationality: Israeli

This boutique dairy is located in Moshav Azaria at the foot of the Jerusalem hills. A local and authentic product line, which is produced by farmers with deep local roots, comprises the base for building values that are based on nostalgia and local connections. The books of Meir Shalev comprise an emotional base for a style of images which create iconographic patterns symbolising a moshav, home, family, and dairy.

伊夫里奶制品

设计机构：Blend-It设计工作室
完成时间：2008年
设计师：阿萨夫·科恩，埃内斯托·比约夫斯基
客户：伊夫里奶制品
国家：以色列

这一乳制品精品店坐落在耶路撒冷小山脚下的莫沙夫·阿扎里亚地区。伊夫里品牌是一个极具地方特色、口味纯正的系列产品，完全由当地土生土长的农民加工，而这也构成了该品牌价值的基础，强调怀旧之情与地方特色之间的结合。梅厄·沙莱夫的图书构成了每个图案的情感基础，每个图案以图示法的设计理念，象征着莫沙夫·阿扎里亚的居民、家乡、家庭和乳制品。

SweetXpressions

Production date: 2008
Designer: Jigisha Patel
Client: SweetXpressions
Nationality: Canadian

This is a selection of coffee cup sleeves designed for SweetXpressions as part of its branding. SweetXpressions specialises in baked goods and chocolate treats.

SweetXpressions烘焙坊

完成时间：2008年
设计师：吉继莎·帕特尔
客户：SweetXpressions烘焙坊
国家：加拿大

这是设计师吉继莎·帕特尔精心为SweetXpressions烘焙坊设计的带杯套的咖啡杯。该项目是SweetXpressions烘焙坊品牌塑造方案的一个部分。SweetXpressions烘焙坊以经营烘焙食品和巧克力食品为特色。

Thermochromic

Production date: 2010
Designer: Elliott Conor Pearson
Client: Personal work
Nationality: British

Needing to have a significant change of colour in order to take advantage of the thermochromic inks, the use of coloured patterns was implemented. The colour change is from somber shades to big and bold colour.

变色杯

完成时间：2010年
设计师：埃利奥特·康纳尔·皮尔森
客户：个人作品
国家：英国

该项目的设计目标是对色彩的运用进行一次伟大的革新，充分利用变色油墨，并设计出异彩纷呈的图案。色彩的变化规律是从暗色调渐变为醒目的色彩。

Le Charme

Production date: 2008
Designer: Petty Hartanto
Client: Le Charme Café
Nationality: Indonesian

Le Charme is a café in Surabaya, which is inspired by Parisian coffee house. The concept is to create a place for rendez-vous and be the centre of meeting place for professionals as well as family dulcet retreats. The contrast between dark chocolate and pastel colours are used to give ambience of the warm feeling of pastries and sweet drinks.

魅力咖啡厅

完成时间：2008年
设计师：佩蒂·哈尔坦图
客户：魅力咖啡厅
国家：印度尼西亚

魅力咖啡厅坐落在印度尼西亚的泗水城，空间的设计灵感源自巴黎的咖啡屋，其设计理念是为上班一族和家庭提供一个约会及放松的完美空间。深色巧克力与柔和彩色的鲜明对比，完美地展现了甜点与甜饮料所传递的温暖之感。

Abaeté - Ybaté

Production date: 2011
Designer: VItor Lopes Leite, Ramon Villain Santos,
Laio de Carvalho, Hellen Aquino Martins
Client: Personal work
Nationality: Brazilian

The project is involving the creation of an ice cream line products for a fictitious brand called Abaeté, and its high quality segment Ybaté - using Brazilian traditional candy as flavours: Brigadeiro (kind of a chocolate truffle), Beijinho (coconut flavoured candy), Cocada (a sweet condensed milk bar with coconut) and Doce de Leite (dulce de leche).

"阿贝蒂" – "雅贝蒂"

完成时间：2011年
设计师：维托尔·洛佩斯·莱特，拉蒙·威廉·桑托斯，莱奥·德·卡瓦略，海伦·阿基诺·马丁斯
客户：个人作品
国家：巴西

该项目是设计师专为一个名为"阿贝蒂"的虚拟品牌下的系列冰淇淋产品及其高端产品"雅贝蒂"提供的设计方案。"雅贝蒂"系列产品拥有若干巴西传统糖果口味，包括布里格德罗（一种松露巧克力）、基西（椰子味的糖果）、可凯达（椰子味甜炼乳）、多西·雷特杜尔塞母乳）。

Mindal Café

Production date: 2010
Designer: Daniil Shumakov
Client: Mindal
Nationality: Russian

Every citizen is looking for its secluded place — to read a book, to dream, to relax with a cup of coffee, to watch the leisurely flow of life from the large windows. Café "Mindal" is a cosy and little coffee shop in town. Corporate identity is based on a pattern that resembles a tablecloth. This conveys a pleasant atmosphere and comfort of home.

敏戴尔咖啡屋

完成时间：2010年
设计师：丹尼尔·舒马克
客户：敏戴尔咖啡屋
国家：俄罗斯

每个人都希望能够拥有一个静谧的地方，在此读书、做梦、享受咖啡所带来的惬意、坐在大大的窗户前闲看花开花落。敏戴尔咖啡屋坐落在一个小镇之上，空间小巧却拥有舒适的氛围。设计师精心打造的企业识别以一个类似桌布的图形为设计基础。这一设计完美地传达出愉悦、如家般的空间感。

La Maison de l'Aubrac

Production date: 2010
Designer: Alexandre Avram
Client: La Maison de l'Aubrac
Nationality: French

This is a cup design for restaurant "La Maison de l'Aubrac" based in Paris. The design uses bright colours to attract cilents' eyes.

"奥贝克的小屋"餐厅

完成时间：2010年
设计师：亚历山大·埃弗拉姆
客户："奥贝克的小屋"餐厅
国家：法国

该项目是设计师亚历山大·埃弗拉姆专为坐落在巴黎的"奥贝克的小屋"餐厅而设计的纸杯方案。鲜明的色调能够轻松捕捉顾客的目光。

Limited Edition Coffee Mug

Design agency: Ph.D-mtl
Production date: 2010
Designer: Phil Héroux
Client: Coffee Depot
Nationality: Canadian

The coffee cup was especially made for the 2010 holiday season, in a limited edition for Coffee Depot, a Canadian coffee shop chain. 600,000 cups were printed and distributed to all Quebec branches. The scarf, as coffee, can keep you warm all winter long.

限量版咖啡杯

设计机构：Ph.D-mtl设计工作室
完成时间：2010年
设计师：菲尔·赫尔科斯
客户：咖啡站
国家：加拿大

咖啡站是加拿大一个咖啡连锁店。这一咖啡杯是设计师菲尔·赫尔科斯专为该连锁店2010年假日季限量版咖啡杯而提供的设计方案。印刷后的600,000个咖啡杯将被分发到所有的魁北克省分店中。其中，温暖的围巾图案令人自然联想起一杯浓香的咖啡在严寒的冬日为人们所带来的丝丝暖意。

Chocobrain

Design agency: Mats Ottdal Design
Production date: 2009
Designer: Mats Ottdal
Client: Chocobrain
Nationality: Norwegian

"Healthy Hot Chocolate on your mind". Chocobrain serves organic Hot Chocolate with low sugar and other drinks to stay or to go. Without any big focus on the name or typeface, the typographic illustration becomes a dominant part of the branding and is the most reconisable identity element.

"印象中的巧克力"咖啡屋

设计机构：马特斯·奥特戴尔设计工作室
完成时间：2009年
设计师：马特斯·奥特戴尔
客户："印象中的巧克力" 咖啡屋
国家：挪威

"印象中的巧克力"，顾名思义，强调人们脑海中的健康热巧克力形象。该品牌咖啡屋以经营有机低糖热巧克力为特色，同时也出售其他饮品，顾客可以根据自身需要在店内享用或打包。在品牌名称或字体没有任何亮点的情形下，字体插图成为这一品牌方案的一个主要部分以及最具识别性的标识元素。

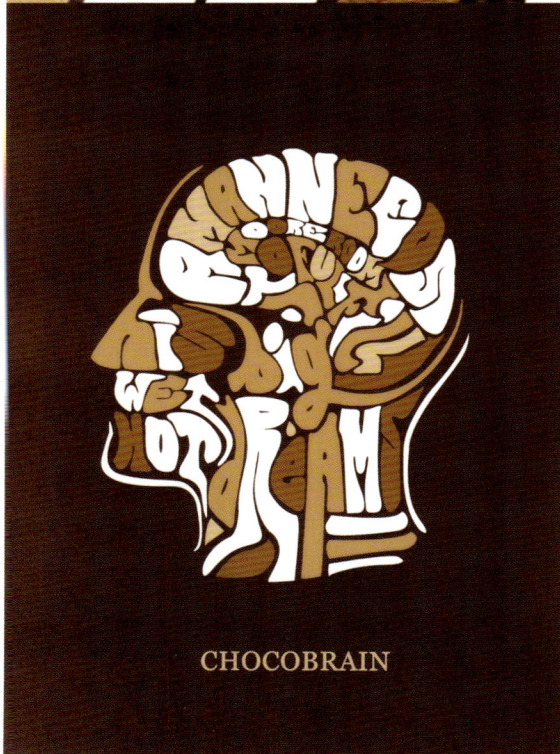

CHOCOBRAIN

Gallery Espresso

Design agency: Savannah College of Art and Design
Production date: 2008
Designer: Jeff Knight
Client: Gallery Espresso
Nationality: American

The re-branding concept for Gallery Espresso evokes calm and stillness with a little flame to keep things hot and spicy. Using colours evoking the autumn season and decorative repeated patterns influenced from India, there is a foreign element that hopes to transport the customer to somewhere peaceful, but slightly unknown.

意式咖啡长廊

设计机构：萨凡纳艺术与设计学院
完成时间：2008年
设计师：杰夫·奈特
客户：意式咖啡长廊
国家：美国

意式咖啡长廊的品牌重塑理念意在利用一个小小火苗唤起宁静、平和之感，同时保留热情与醇香的特质。设计师精心选用的配色方案令人自然联想起硕果累累的秋季，而重复的装饰图案也悄悄地流露出印度的气息，设计师力图运用一个舶来元素为顾客营造出一个祥和而又从未体验过的空间氛围。

Coffee Time

Design agency: Studio43
Production date: 2008
Designer: Sergio Laskin
Client: Coffee Time
Nationality: Latvian/Russian

Riga-based coffee shop — Coffee Time brings us to 20's with retro designs and tasty coffee (as well as teas, milkshakes, sandwiches, etc.), creating a very warm and unique atmosphere!

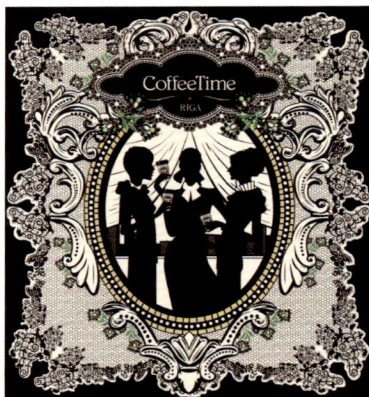

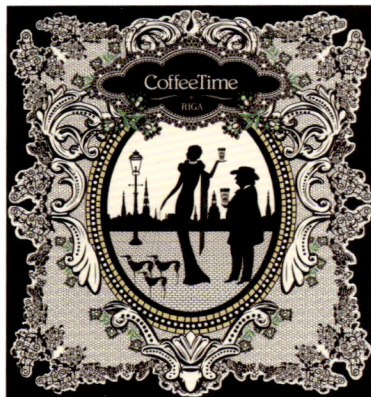

咖啡时光

设计机构：43号设计工作室
完成时间：2008年
设计师：塞尔吉奥·拉斯金
客户：咖啡时光
国家：拉脱维亚，俄罗斯

坐落在里加的"咖啡时光"咖啡馆将引领顾客进入到一个拥有20世纪20年代设计风格并出售美味咖啡（同时也出售茶、奶昔、三明治等）的美妙地带，使顾客自由自在地在此享受到温馨、独特的环境。

"Igloo"- Ice Cream on the Rocks

Production date: 2010
Designer: Yotam Bezalel
Client: "Igloo"- Ice Cream on the Rocks
Nationality: Israeli

Igloo is an ice cream chain, brought to Israel for the first time. Its unique method of custom-made ice cream affords the customer a fun and colourful experience. Branding the chain was first done by designing the logo which reflects the values of the brand. Then combining a clean design along with the development of an original retro style font corresponding to the ice cream shape. The graphic language intensifies the experience with the use of fun and colourful icons. The project included full image design, strategy, copy-writing, design of the logo, packaging, signs, website, etc. The project was done in collaboration with Anna Geslev studio.

"冰屋"-岩石上的冰淇淋

完成时间：2010年
设计师：约塔姆·比扎莱尔
客户："冰屋"-岩石上的冰淇淋
国家：以色列

"冰屋"是首次进入到以色列的连锁冰淇淋店。经典、独特的冰淇淋制作手法为顾客提供了一个妙趣横生、色彩缤纷的感官体验。该品牌的塑造方案首先从标识的设计入手，旨在彰显出该品牌的价值。随后，设计师巧妙地将一个简洁的设计风格与一个改进的复古风格字体完美结合，从而与冰淇淋的外形相得益彰。独特的平面语言强化了诙谐、彩色图标的运用。该项目包括所有图案、设计战略、文字的复制、标识、包装、标志、网站的方案设计。该项目由设计师约塔姆·比扎莱尔与安娜·杰斯雷夫品牌设计工作室合力打造。

Gaufres & Goods

Design agency: Andrew Deming Design
Production date: 2009
Designer: Andrew Deming
Client: Gaufres & Goods, St. Augustine, FL
Nationality: American

Gaufres & Goods is a beautifully warm, one-of-a-kind European restaurant located in downtown St. Augustine, FL. The design and application of the identity throughout the various materials lends Gaufres & Goods the legitimacy and refinement that it previously lacked in its design, while incorporating through textile patterns, the warmth one feels when dining in.

Gaufres & Goods餐厅

设计机构：安德鲁·戴明设计工作室
完成时间：2009年
设计师：安德鲁·戴明
客户端：佛罗里达州圣奥古斯丁Gaufres & Goods餐厅
国家：美国

Gaufres & Goods餐厅是一个美妙、温馨的欧式餐厅空间，坐落在美国佛罗里达州圣奥古斯丁的市中心。全新品牌识别系统的设计方案与应用遍及所有的宣传材料，巧妙地传达出餐厅原有设计所无法彰显的正统与精致之感，同时结合纺织图案，流露出就餐环境的温馨与浪漫。

Jasmine Café

Design agency: Manara Design Studio
Production date: 2011
Designer: Amer Amin
Client: Jasmine Café
Nationality: Palestinian

Jasmine Café is the first Palestinian café to produce their own coffee, so as the creative workforce the designer had to come up with something ideal and unique for this project, hence the yellow and orange cup, which represents the natural colours of the Palestinian Jasmine flower, while brown being the natural colour of roasted coffee beans. The pattern represents the continous fresh coffee taste.

茉莉咖啡厅

设计机构：马纳拉设计工作室
完成时间：2011年
设计师：阿米尔·阿明
客户：茉莉咖啡厅
国家：巴勒斯坦

茉莉咖啡厅是巴勒斯坦首家自己加工咖啡并出售的咖啡零售空间，因此，担任该品牌创意设计任务的设计师阿米尔·阿明力图为该项目打造出一个完美而独特的设计方案。黄色和橘色最终被选为纸杯的主色，代表着巴勒斯坦茉莉花的自然色彩，而棕色则象征着烘焙咖啡豆的自然色调。除此之外，纸杯上的图案象征着醇厚、绵长的咖啡味道。

MUG Beer

Production date: 2011
Designer: Ivan Maximov
Client: MUG Pub
Nationality: Russian

This is a new concept for take-away beer. Beer is filled into paper cups and put into a carrier. A sticker is placed on the lid to identify the brew as well as the date it was filled. The new cup combines the form of the traditional beer pint and usability of recyclable paper cups.

马克啤酒

完成时间：2011年
设计师：伊凡·马希莫夫
客户：马克酒吧
国家：俄罗斯

该项目是便携式啤酒的一个全新包装概念。装有啤酒的纸杯经封装之后被置放在一个手拎式包装盒内。除此之外，纸杯盖子上的粘贴有效地突出了啤酒的品牌以及灌装日期。在此，新式纸杯与传统的啤酒印刷模式以及极易回收价值的纸杯材质完美结合。

Délice Ice Cream

Production date: 2011
Designer: Bo Mouridsen, Kasper Andersen
Client: Délice Ice Cream
Nationality: Danish

A simple cup design for quality ice cream. The base colour is black to give it a classy look, and the flavours are divided by colours (brown is chocolate, white vanilla, etc.). The cups are made in two sizes, a single-size and a family-size. The name is written in front and on top, for an easy overview.

快乐冰淇淋

完成时间：2011年
设计师：波·莫里德森，卡斯帕·安徒生
客户：快乐冰淇淋
国家：丹麦

这是设计师波·莫里德森与卡斯帕·安徒生专为一款优质冰淇淋而设计的简约风格纸杯。整个纸杯的基础色调为黑色，外观优雅，其内部盛放的不同冰淇淋口味也同样按照颜色进行区分。这一系列纸杯分成两种型号，其中一个是单杯装，另一个则是家庭装。该品牌冰淇淋的名称被印刷在纸杯的前端和顶部，便于识别和捕捉观者的目光。

Kavalkade Coffee Tumblers

Production date: 2009
Designer: Katrine Austgulen
Client: Kavalkade Coffee Tumblers
Nationality: Norwegian

The coffee tumblers represent India and the gothic, classical and renaissance era. The design is inspired by paintings, architecture and typography from the time periods, and the culture of India. The integrated window offers one a glimpse into the logo, which has a mixed look that combines the old with the contemporary.

卡瓦尔卡德咖啡杯

完成时间：2009年
设计师：凯特琳·奥斯特格伦
客户：卡瓦尔卡德咖啡
国家：挪威

卡瓦尔卡德咖啡杯的设计流露出浓厚的印度气息，象征着哥特、古典与文艺复兴时代风格的完美融合。设计师凯特琳·奥斯特格伦取材自以上几个时期的画作、建筑和字体设计风格以及印度的传统文化。这一系列纸杯中均没有类似的窗口，便于使其中设置的标识能够轻松捕获观者的目光，而标识的设计则遵循了古典与现代相结合的原则。

Molloy's Bakery & Coffee

Production date: 2010
Designer: Olly Blake
Client: Molloy's Bakery & Coffee Shop
Nationality: Irish

A staple of Bray in Co. Wicklow, Molloy's has been around for 50 years. The designer introduced folk tales of characters and features of Bray from Victorian times when Bray was a seaside resort into the identity. He screen printed cups, place mats, menus and aprons among other things.

莫洛伊的面包&咖啡店

完成时间：2010年
设计师：奥利·布雷克
客户：莫洛伊的面包&咖啡店
国家：爱尔兰

在爱尔兰威克洛市布赖地区占绝对市场份额的莫洛伊的面包&咖啡店已经拥有50多年的开办历史。设计师奥利·布雷克巧妙地将维多利亚时代布赖作为完美海滨度假胜地的民间故事人物和特色引入到识别设计之中。除此之外，对于纸杯、餐垫、菜单和围裙等图案设计，设计师专门采用了丝网印刷的手法。

Wi-Fi Coffee Cup

Production date: 2010
Designer: Diyana Nikolova
Client: Wi-Fi Coffee Bar
Nationality: Italian/Bulgarian

The project Wi-Fi Coffee Bar is for younger target. It describes brand identity for a possible American-English style coffee bar with free wi-fi. The logo marries elements that evoke more to the coffee with the technology in a vintage ambente in order to create a warm athmosphere but also modern and youthful.

无线网络咖啡杯

完成时间：2010年
设计师：戴安娜·尼可洛娃
客户：无线网络咖啡吧
国家：意大利/保加利亚

无线网络咖啡吧以年轻群体为消费对象。该项目旨在为这一洋溢着英美混合气息并提供免费无线网络的咖啡吧设计一个品牌识别系统。精致的标识与体现这一咖啡吧"复古环境中的高科技"的元素完美搭配，共同营造出一个温馨、时尚而充满活力的空间氛围。

United Bakeries

Production date: 2006
Designer: Sukker Design
Client: Sukker Design
Nationality: Norwegian

This is a development of United Bakeries brand including packaging and the development of several product brands.

联合面包店

完成时间：2006年
设计师：苏克尔设计工作室
客户：苏克尔设计工作室
国家：挪威

该项目是苏克尔设计工作室专为联合面包店而提供的品牌开发方案。该方案包括产品的包装和若干产品的品牌开发。

Pinkberry

Design agency: Ferroconcrete
Production date: 2005
Designer: Owen Gee, Priscilla Jimenez, Ann Kim, Sunjoo Park, Wendy Thai
Photography: Vanessa Stump
Client: Pinkberry
Nationality: American

粉红浆果酸乳店

设计机构：钢筋混凝土设计工作室
完成时间：2005年
设计师：欧文·吉吉，普里西拉·希门尼斯，安·金姆，尚约·帕克，温迪·泰
摄影师：凡妮莎·斯达姆
客户：粉红浆果酸乳店
国家：美国

Customers stand in line for a taste of the yogurt, but they also get in that line because of the brand. Light, sweet but not too sweet, fresh pops of colour or a modern white backdrop, minimal yet infused with Swirly Goodness – that describes not only the yogurt, but also the website, the packaging, the menus, the marketing materials, the stores, the newsletters, and even the gift cards. Ferroconcrete fell in love with the essence of Pinkberry and, inspired by the yogurt and the modern décor of the first store, created an entire personality for customers to love.

顾客排队购买一种口味的酸奶，其实也是因为受到了这一品牌的吸引。轻盈、甜美而不做作，清晰、时尚的色调搭配流行的白色底色，完美地传递出简约之感，此外，"善意的谎言"这一标语被广泛地应用到酸奶、网站、包装、菜单、营销材料、店铺、简讯乃至礼品卡片之上。钢筋混凝土设计工作室钟爱"粉红浆果"这一品牌的纯粹之美，于是从酸奶和该品牌第一家店面的现代化装饰风格中寻找灵感，打造了这个极具个性化的设计项目，从而深得顾客的喜爱。

pinkberry®

SWIRLY GRAM

pinkberry®

YOU MAKE MY
HEART SWIRL
HAPPY VALENTINE'S DAY

50% Café Cup

Design agency: JD Studio
Production date: 2010
Designer: José Diego
Client: 50% Café
Nationality: Spainish

The designer would like to highlight the simple design. Blue can give people a sense of calm. Blue and white mix, you can let us think of sky and clouds. When you get to drink this cup of coffee, the design will provide an easy feeling around you.

"50%咖啡厅" 纸杯

设计机构：JD设计工作室
完成时间：2010年
设计师：何塞·迭戈
客户："50%咖啡厅"
国家：西班牙

来自JD设计工作室的设计师何塞·迭戈希望能够突出设计的简约之美。蓝色，能够在视觉上给人带来一种宁静、淡然之感。而蓝色和白色的混合搭配则能够让人自然联想起天空和白云。如此设计，旨在于为正在品尝这一咖啡的顾客营造出一种轻松、悠闲的气息。

Café Muffin — Chain Cafés

Production date: 2010
Designer: Ewelina Bocian
Client: Chain Cafés
Nationality: Poland

More than 60% of chain cafés clients are woman — referring to the statistics. Café muffin is chain cafés addressed to woman in all ages. Project focuses on the take-away packaging for coffee and sweets, not only for big amounts of sweets, but also just the symbolic ones — like the feminine and elegant box for 4 chocolates or a single muffin box. All designed with elegance that no woman could resist.

松饼咖啡厅—连锁咖啡厅

完成时间：2010年
设计师：艾维里那·波西恩
客户：连锁咖啡厅
国家：波兰

根据数据显示，连锁咖啡厅60%以上的顾客均为女性。松饼咖啡厅，这一连锁咖啡厅，以各年龄段的女性为消费对象。该项目以咖啡和糖果的便携式包装设计为重点，除了大量的糖果包装之外，同时也包括一些极具象征意义的包装，例如柔美、优雅的四颗巧克力包装盒或一个单只小松饼盒。所有的设计均完美地散发出高贵、端庄之气息，令无数女士们无法抗拒。

Our Cartoon Christmas

Production date: 2010
Designer: Duncan Jones
Nationality: New Zealand

There's a child living in every adult's mind. These Cartoon Christmas paper cups intend to awaken our memory of happy childhood during Christmas holidays. The colourful images of Christmas make us immersed in a festival atmosphere.

"我们的卡通圣诞节"

完成时间：2010年
设计师：邓肯·琼斯
国家：新西兰

每个人的脑海中都住着一个孩子。这一圣诞节纸杯旨在唤醒人们对儿时的回忆，想象圣诞节即将到来时的喜悦之情。丰富的色调和多姿多彩的圣诞节图像为整个设计增添了许多节日气氛。

Graphic Cup

Design agency: band-design
Production date: 2011
Designer: George Hartanto
Nationality: American

Cup is everywhere and everyday in our life. The successful design of paper cup will make our ordinary life filled with different colours. Point and line are the basic elements of design, and this series of paper cups can be classified by different designs of dot, line and area. The designer, George Hartanto, uses the idea of point and line in his other works as well. Thus, George combines simple patterns and rich colours in this project, and hopes these cups will colour our unremarkable life.

图形纸杯

设计机构：团队设计工作室
完成时间：2011年
设计师：乔治·哈尔坦托
国家：美国

杯子在我们的生活中无处不在。在人们的日常生活中，几乎每一天都会被用到。一个成功的纸杯设计将为人们苍白的生活增添许多色彩。点、线是该项目最基本的设计元素，这一系列纸杯的设计完全可以根据点、线、面进行区分。这些元素也经常出现在设计师乔治·哈尔坦托的其他作品之中。因此，在该项目中，设计师巧妙运用简约的图形并搭配鲜明的色调。设计师希望这一设计能够为人们的平凡生活添加些许色彩。

Dunkin' Donuts

Design agency: Sterling Brands
Production date: 2009
Designer: Irina Ivanova
Client: Dunkin' Donuts
Nationality: American

The cup is designed for the 60th anniversary of the Dunkin' Donuts. Texts and graphics as the basic elements combine together to compose a pattern. Specially, the texts are all variations from the font of Dunkin' Donuts. The striking "Happy 60th Birthday" could easily catch the eyes of customers and also highlight the theme of design.

唐恩都乐咖啡文化馆

设计机构：斯特林品牌设计工作室
完成时间：2009年
设计师：伊琳娜·伊万诺娃
客户：唐恩都乐咖啡文化馆
国家：美国

这个纸杯是为了唐恩都乐咖啡文化馆 60周年纪念而设计的。这个设计主要以文字和图形结合，构成整个设计的图案。上面的文字主要是各种字体的唐恩都乐咖啡文化馆的变形。醒目的60周年快乐，当人们看见这个纸杯时首先映入眼帘，突出了主题。

Childhood Memory

Design agency: BD
Production date: 2011
Designer: Damon Scott
Nationality: British

The inspiration of this project comes from the designer Damon Scott's memory of his mother's knitting in his childhood. So Scott means to remind people of their cheerful memory and joyful recall at Christmas in their childhood.

童年的记忆

设计机构：BD设计工作室
完成时间：2011年
设计师：达蒙·斯科特
国家：英国

该项目的设计灵感源自设计师达蒙·斯科特对儿时妈妈为其编织毛衣的回忆。设计师力图通过这一设计唤起人们对儿时圣诞节快乐场景的回味与追忆。

Hot Cup

Production date: 2011
Designer: Hyman Nikolova
Nationality: British

The cups are designed for a gift shop. The shop gives people a young feeling. So the designer made this small design. Pure, fresh, and young. When the guest buys things, the shop owner brings out a cup of warm coffee. The customer will be also very warm.

热饮杯

完成时间：2011年
设计师：海曼·尼可洛瓦
国家：英国

这一系列纸杯由设计师海曼·尼可洛瓦专为一个礼品店而设计。精致的店面设计给人留下了朝气蓬勃的印象。因此，设计师海曼·尼可洛瓦也同样为这个小型设计方案添加了清纯、清晰、活力四射的气息。当店主为进到店内的顾客端来一杯温热的咖啡之时，顾客也势必被这种温馨的氛围所感染。

Colour's Hotcup

Design agency: Design Life
Production date: 2011
Designer: Jacqueline Edwina
Nationality: British

Life needs colour, which is also the designer Jacqueline Edwina's design philosophy. The design inspiration of colourful patterns is derived from daily life, and in the meantime, brings different tastes to tasteless life.

彩色的热饮杯

设计机构：设计生活工作室
完成时间：2011年
设计师：杰奎琳·埃德温娜
国家：英国

生活需要色彩。这也是设计师杰奎琳·埃德温娜设计这一纸杯的意图所在。构成整个图案的彩色图形正如同生活中的点点滴滴。设计源自生活，同时又为平淡的生活增添许多色彩。

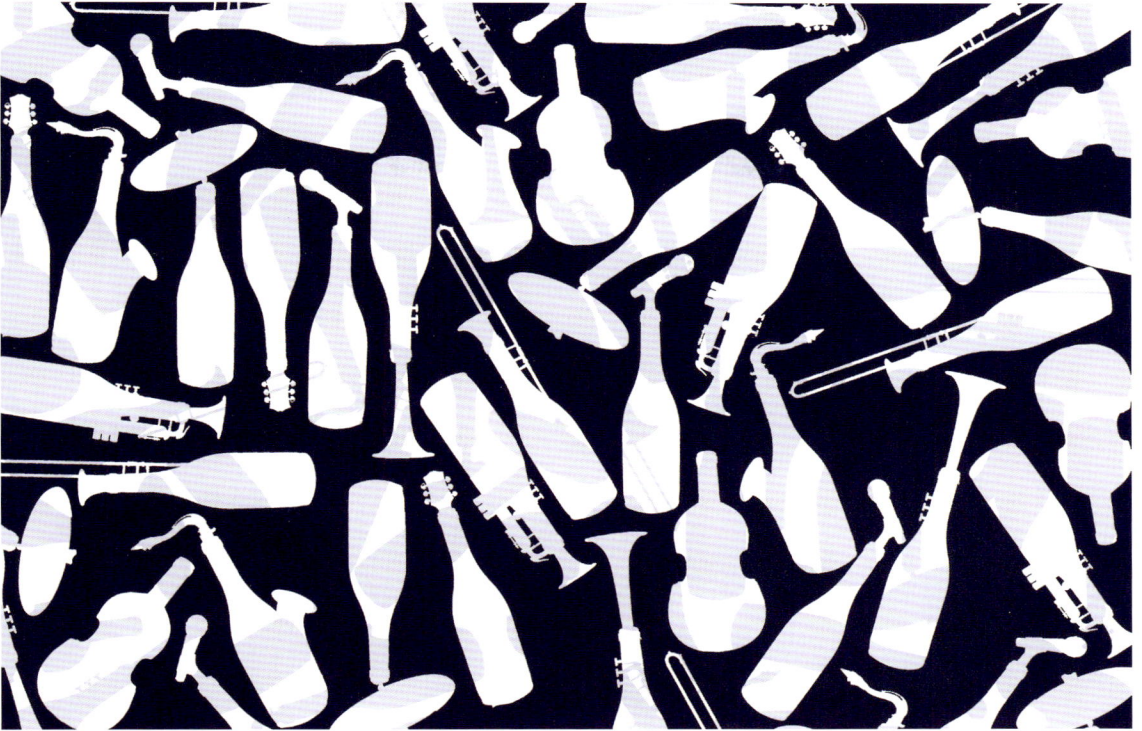

The Parlour Paper Cup

Production date: 2011
Designer: Charlie Nolan
Nationality: British

The task was to create a brand and design promotional material for a delicatessen named "The Parlour". The brand is based on jazz music. By day the deli served local and homemade products and by night it transformed into an upper class restaurant. Charlie Nolan designed a logo, business cards, bags, sandwich bags, disposable coffee cups, posters, fruit juices and other promotional material which would be used in the deli/restaurant. Applications used included screen printing, letterpress, laser cutting, Adobe Photoshop and Illustrator.

帕勒纸杯

完成时间：2011年
设计师：查理·诺兰
国家：英国

这个项目主要是为"帕勒"食品店设计商标和宣传资料。设计商标的灵感来自于爵士乐。 白天，食品店供应当地手工制作的食物，然而到了晚上，食品店则变身为高级餐厅。设计师设计了可在食品店或餐厅使用的商标、名片、袋子、三明治袋子、一次性咖啡杯、果汁杯、海报和其他宣传材料。设计师使用的工具包括丝网印刷、凸版印刷、镭射切割、图像处理软件和矢量图形软件。

Idea Cup

Design agency: MIN Design Studio
Production date: 2011
Designer: Amer Amin
Client: Energy Factory
Nationality: Palestinian

Idea Cup is designed for Energy Factory. This factory needs a concise industrial feeling. So the designer created two patterns, a square one and a circular one, both with a rough background to give a strong industrial feeling.

概念纸杯

设计机构：MIN设计工作室
完成时间：2011年
设计师：阿米尔·阿明
客户：能源工厂
国家：巴勒斯坦

概念纸杯由设计师阿米尔·阿明专为能源工厂而设计。这一工厂希望设计能够精确地传达出工业之感。因此，设计师根据客户的要求精心设计出两款图案，即简约的方形和圆形，并以粗糙的界面为背景。如此设计，旨在传递给顾客以刚劲的工业之感。

Solberg & Hansen Coffee

Design agency: Anti
Production date: 2009
Designer: Fredrik Melby & Martin Stousland
Client: Solberg & Hansen Coffee
Nationality: Norwegian

The brand, logo and the signature symbol along with typography and the precise information about coffee help the brand step up its position in the market and stand out. By adding the fresh blue colour together with the ability to tell a story through careful and well thought-out design, this new image shows that this brand is urgently trying to differentiate and stand out from the traditional competitors that exist in this market.

索伯格&汉森咖啡厅

设计机构：安蒂设计工作室
完成时间：2009年
设计师：弗雷德里克·梅尔比，马丁·斯图思兰德
客户：索伯格&汉森咖啡厅
国家：挪威

品牌、标识、带有特殊设计字体的文字、有关这一品牌咖啡的准确信息为该品牌建立市场定位并脱颖而出奠定了基础。设计师弗雷德里克·梅尔比与马丁·斯图思兰德通过将亮蓝色与细致考究的图像完美结合，进而将这一品牌力图独树一帜的进取之心充分地彰显出来，并将该咖啡厅和市场上传统的竞争对手拉开距离。

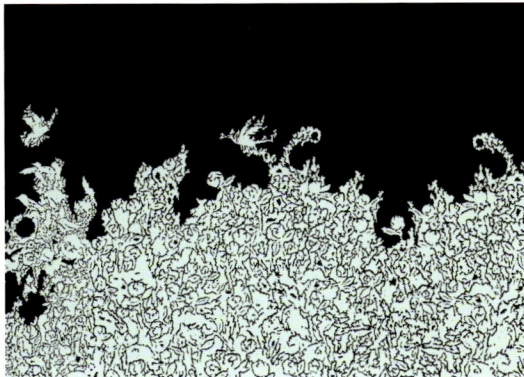

Rin Green Tea

Design agency: Tinytwiggette Design
Production date: 2011
Designer: Lynn Nguyen
Nationality: Australian

This completely adorable green tea packaging for individual teabags is targeted toward young adult females (13–25). The aim of this project is to promote the healthy benefits of green tea and to capture the attention of the female audience with irresistible cute designs. Each flavour has been characterised with a unique character and distinctive wallpaper design. The structure of the package reflects the paper coffee/tea cup that will definitely enhance the shelf appeal and presentation. The cup design is also manufactured from a 100% renewable resource that is 100% recycle & biodegradable that may enhance the brand values of the product. Rin Green Tea branding also aims to communicate the audience's environmental commitments. Young female audience will not be able to resist the cuteness that oozes out of these designs.

里恩绿茶

设计机构：Tinytwiggette设计工作
室
完成时间：2011年
设计师：阮林恩
国家：澳大利亚

这一专为独立绿茶包而设计的包装造型可爱
而独特，主要面向13~25岁之间的女性消费群
体。该项目的设计目的是有效地宣传绿茶对身
体的有益之处，并通过魅力难挡的精致设计成
功捕获女性消费主体的目光。每种口味均被赋
予了一个独特的人物和壁纸造型。匠心独运的
包装结构决定了这一系列纸质咖啡杯或茶杯在
货架上必然独树一帜、力压群雄。该系列纸杯
以100%可再生资源为加工原料，可完全回收
及生物降解，有效地增加了这一产品的品牌价
值。里恩绿茶品牌塑造的目的在于充分地向顾
客传达出其真诚的环保理念。这些可爱、细腻
的设计风格相信每位年轻女士都无法抗拒。

Moomah

Design agency: Apartment One
Production date: 2011
Designer: Spencer Bagley
Client: Tracey Stewart
Nationality: American

The designers created three different variations of the M logo out of elements that visually represented each of the four core brand values: connect, create, discover, and nourish. They also wanted to create a memorable and recognisable brand language without having to put a logo on all applications. They created custom, application-specific illustrations filled with the wonder, magic, and heart of Moomah and placed them above a dotted line and type to set the foundation for the brand language. These illustrations and corresponding tag lines were conceptually developed for their specific applications.

莫麦哈咖啡厅

设计机构：一号公寓设计工作室
完成时间：2011年
设计师：斯宾塞·巴格利
客户：特蕾西·斯图尔特
国家：美国

设计师斯宾塞·巴格利等人利用若干能够在视觉上传达出品牌价值四个核心的（联合、创建、发现与培养）元素创建三个不同的标识变体。除此之外，设计师也力图打造一个极具感染力和识别性的品牌语言，无需在所有的应用程序中添加标识。设计师专为这一品牌精心打造了一系列专用插画，并赋予其奇妙、神奇、莫麦哈咖啡厅的真诚之感，随后将其置放在一条虚线以及文字的上方，从而奠定这一品牌语言的基础。这些插画以及相应口号的概念性开发全部以它们的特定应用为出发点。

sip slowly

Cat in the Cup

Production date: 2011
Designer: Jenna Taylor
Client: Personal work
Nationality: American

This special edition line of coffee cups was created to promote the cinematic release of Dr. Seuss cartoons. Line includes coffee cup sleeves inspired by Thing 1 and Thing 2, as well as a coffee cup referencing Cat in the Hat.

杯子中的猫

完成时间：2011年
设计师：珍娜·泰勒
客户：个人作品
国家：美国

这一特别版系列咖啡杯专为宣传改编自苏斯博士漫画的动画电影而设计。这一系列咖啡杯杯套的设计源自设计师对苏斯博士漫画 "Thing 1 &Thing 2" 的参考，而咖啡杯的设计则取材自漫画 "帽子里的猫"。

Peppes Pizza

Design agency: Sukker Design
Production date: 2011
Designer: Gro Vik
Client: Umoe Restaurant Group AS
Nationality: Norwegian

This is a complete redesign of the largest Pizza restaurant chain in Scandinavia. The designers developed the full brand experience from conceptual development, identity, packaging to interior design.

派乐士比萨

设计机构：苏克尔设计工作室
完成时间：2011年
设计师：格罗·维克
客户：Umoe餐厅集团
国家：挪威

该项目是设计师专为斯堪的纳维亚半岛最大的连锁比萨餐厅而设计的品牌重塑方案。设计师格罗·维克等人开发的这一系统化的品牌方案包括品牌概念的开发、识别、包装以及室内设计。

Oello

Design agency: Noote&Netoo
Production date: 2009
Designer: Yuliana
Client: Oello
Nationality: Indonesian

Frozen yoghurt brings out the childlike innocence in all of us. The designers picked the name from Incan goddess who taught the art of spinning. The brand hopes to celebrate the spirit of fun, tasty & freshness.

奥艾罗酸奶

设计机构：Noote&Netoo设计工作室
完成时间：2009年
设计师：尤利亚纳
客户：奥艾罗酸奶
国家：印度尼西亚

冷冻酸奶能够唤起人们的童心。来自Noote&Netoo设计工作室的设计师从传授纺纱艺术的印加女神中找到灵感并为该品牌进行命名。这一品牌设计方案力图体现出其风趣、美味及清新的特色。

YOYO – Frozen Yogurt Chain

Design agency: Adlai & Partners Ltd.
Production date: 2008
Designer: Adlai & Partners Ltd.
Client: YOYO - Yogurt Land
Nationality: Israeli

The success of Frozen Yogurt chains in the United States such as Pinkberry and Red Mango, initiated the trend in Israel. The designers came up with the name "YOYO" and created a cute brand character — a distant relative of a penguin with a yoyo swirl on his head... YOYO is all about fun and wellness without compromising taste.

"溜溜球" – 冷冻酸奶连锁店

设计机构：阿德莱合作伙伴有限公司司
完成时间：2008年
设计师：阿德莱合作伙伴有限公司
客户：溜溜球 – 酸奶地带
国家：以色列

美国连锁冷冻酸奶店，如粉红浆果和红芒果的成功开办激发了以色列酸奶店的创办热情。阿德莱合作伙伴有限公司经过精心的调查与研究，最终确立了"溜溜球"这一名称，并创建了一个可爱的品牌形象，图片中的南极企鹅憨态可掬，其头上还带有一个漩涡状溜溜球。"溜溜球"冷冻酸奶的经营理念是在确保口感丰富的同时，不失风趣和健康之美。

143

Brew Nerds

Design agency: Mitre Agency
Production date: 2007
Designer: Ryan McCullah
Client: BrewNerds Coffee Shop

At a certain point during the pursuit of a hobby or interest you crossover from being an expert to a full-on nerd. That's exactly what happened with Brew Nerds. Looking beyond the fluff and snobbery of modern coffee, Brew Nerds applies actual science and technology to formulate blends and concoct brews calibrated to absolute perfection. The designer drew this logo and supporting packaging doodles shortly after his first meeting with Brew Nerds in a moment of channeling his inner nerd (there's one in all of us). Embrace the nerd inside you!

"书呆子"咖啡厅

设计机构：米特芮设计公司
完成时间：2007年
设计师：瑞安·麦克库拉
客户："书呆子"咖啡厅

在追求某一业余爱好或兴趣的问题上，专家和书呆子之间的距离仅一步之遥。而这一切正在"书呆子"咖啡厅中上演。不同于那些吹毛求疵、势力浮华的现代咖啡厅，"书呆子"咖啡厅应用现行的科学和技术冲泡和调和咖啡，从而加工出经典极致的咖啡。设计师瑞安·麦克库拉在进入这一咖啡厅的瞬间，其心中的那个"书呆子"（每个人心中都住着一个"书呆子"）瞬间被激活，随后绘制出这一咖啡厅的标识和相关的包装涂鸦。拥抱你内心中的那个"书呆子"吧！

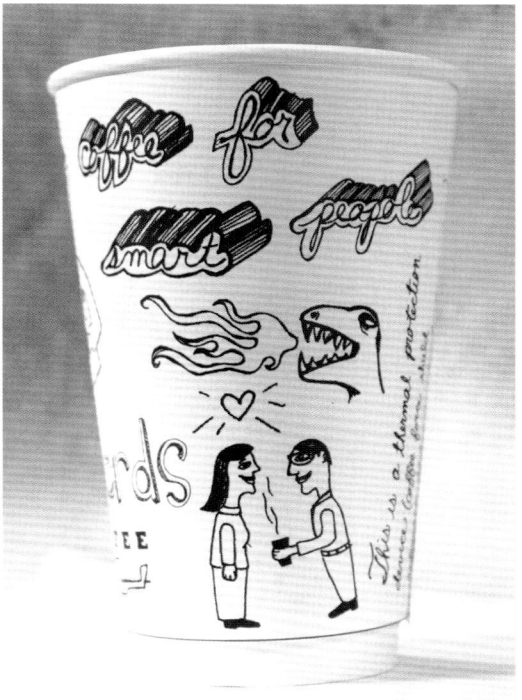

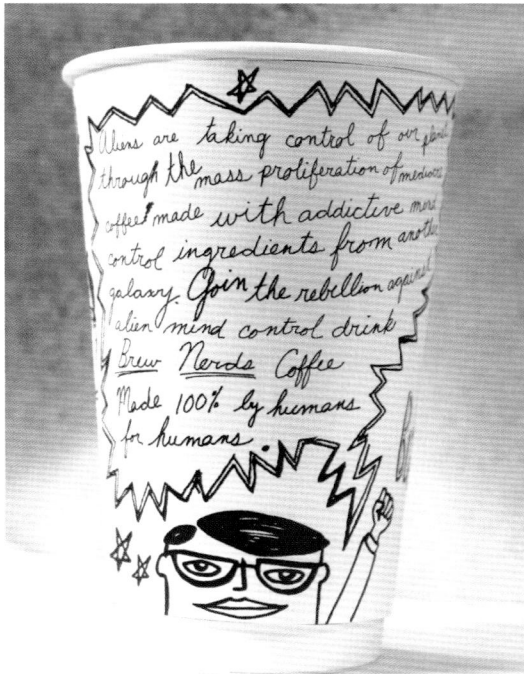

CSM (Central Saint Martins)

Design agency: CSM (Central Saint Martins)
Production date: 2008
Designer: Isabel Eeles
Client: Potential Brick Lane Coffee Shops
Nationality: British

The design is a set of 5 limited edition collectible paper cup exteriors inspired by Brick Lane to infuse flair into the local café scene. The design contains many influences of the area collated and combined to create a mismatch of components reflecting the overwhelming impact of the infamous street, full of colour, conflicting geometry and life. In a font inspired by existing graffiti, "Brick Lane" runs down the centre tying all 5 pieces together. It is divided in a jigsaw puzzle style with angular pieces to help ease assembly of the full design encouraging customers to buy, keep, collect them all to reveal the final image.

圣马丁中央艺术与设计学院

设计机构：圣马丁中央艺术与设计学院
完成时间：2008年
设计师：伊莎贝尔·艾乐斯
客户：极具潜质的布里克巷咖啡店
国家：英国

设计五个为一套的限量版收藏纸杯，其外部的图案设计灵感源自布里克巷，旨在将独特的设计风格融入到当地咖啡厅之中。设计将该地区许多极具影响力的事物进行整理和结合，从而打造了一个不协调的组合，进而彰显出一个声名狼藉的大街的强大影响力。丰富的色调，与几何学和生活形成强烈的碰撞。就字体来说，设计师取材自现有的涂鸦，巧妙地将"Brick Lane"（布里克巷）中的每个英文字母设置在杯子的中央，从而使这五个部分建立起完美的衔接。除此之外，设计师利用直角部分巧妙地将图案划分成一个七巧板的模式，以降低整套设计收集和匹配的难度，从而鼓舞消费者购买、收藏，最终看到一个整体形象。

Brick Lane in a Cup

纸杯中的布里克巷

Production date: 2006
Designer: Priz Praz Pruz
Client: Central St Martins College of Art and Design
Nationality: Paraguayan

完成时间：2006年
设计师：普利兹·普拉兹·普鲁兹
客户：圣马丁中央艺术与设计学院
国家：巴拉圭

These are customised coffee cups for art foundation project in Central St Martins College of Art and Design.

这一定制的系列咖啡杯是设计师专为圣马丁中央艺术与设计学院艺术基金会项目而设计。

Clouds 1 and 2

Production date: 2010
Designer: Seth Beukes
Client: Personal work
Nationality: South African

This project consists of hand-drawn art on paper cups done with a fine line pen. In this series the designer experimented with cloud patterns. Clouds 1 feathers an eye as the central focus. Clouds 2 has a moon as the main focus which was given with a coffee stain.

两张云朵图

完成时间：2010年
设计师：赛斯·贝克斯
客户：个人作品
国家：南非

该项目指的是运用细线笔在一系列纸杯上进行手绘艺术创作。在这一系列纸杯中，设计师赛斯·贝克斯尝试绘制了云朵的图案。其中，第一个图案以云彩中的眼睛为焦点。第二个云朵图案则以满月为重点，而满月完全是利用咖啡渍而设计。

Cavistons Coffee Cup

Design agency: Steve Simpson Illustration & Design
Production date: 2011
Designer: Steve Simpson
Client: Cavistons Food Emporium
Nationality: Irish

Cavistons is a fabulous deli situated in a coastal village just outside Dublin. Steve Simpson was asked to create a design for a disposable coffee cup. The only brief was to make it fun and eye-catching. The printed cups are double walled with a wonderfully tactile matt finish.

Cavistons 咖啡杯设计

设计机构：史蒂夫·辛普森插画与设计工作室
完成时间：2011年
设计师：史蒂夫·辛普森
客户：Cavistons 食品百货商场
国家：爱尔兰

Cavistons 是一家极好的熟食店，坐落在都柏林郊外的海滨村庄里。该熟食店邀请史蒂夫·辛普森设计工作室设计一款一次性咖啡杯，他们唯一的要求便是这款设计要非常的有趣并且吸引眼球。印花的白字采用双层杯壁设计，有着奇妙的亚光触感。

Ozone

Design agency: TGM Design
Production date: 2009
Designer: Craig Jones
Client: Ozone Coffee Factory
Nationality: New Zealand

Ozone Coffee Factory selected a unique design to promote their Fair-trade Coffee. Featuring a large coffee plant illustration and bold coffee cherries, this design was a real success.

新鲜空气地带

设计机构：TGM设计工作室
完成时间：2009年
设计师：克雷格·琼斯
客户：新鲜空气地带咖啡工厂
国家：新西兰

新鲜空气地带咖啡工厂选择了一款独特的设计来促进其咖啡的销售。该设计以一颗大的咖啡树的插画和粗大的咖啡樱桃为特色，这是一款真正成功的设计。

JJ Bean Coffee

Design agency: Thought Shop Creative Inc.
Production date: 2009
Designer: Kristin Hubbard
Client: JJ Bean Coffee
Nationality: Canadian

This is complete overhaul of the inventory of JJ Bean's printed materials and corporate story. The work included crafting a series of custom illustrations depicting their East Vancouver roots, as well as all point of communications between customers and JJ Bean.

"JJ咖啡豆"

设计机构：思维店铺创意公司
完成时间：2009年
设计师：克里斯汀·哈伯德
客户："JJ咖啡豆"咖啡厅
国家：加拿大

该项目是设计师克里斯汀·哈伯德对"JJ咖啡豆"咖啡厅所有印刷材料与企业故事的一次全新改造。这一工程包括一系列描述该公司东温哥华发源地的特制插画、所有和顾客与"JJ咖啡豆"咖啡厅互动的相关材料的设计。

Coffee Cup Faces

Production date: 2010
Designer: Jess Giambroni
Client: Personal work
Nationality: American

表情咖啡杯

完成时间：2010年
设计师：杰西·吉亚姆布罗尼
客户：个人作品
国家：美国

The designer likes to drink coffee and usually visits a café across the street from the design studio where she works in San Francisco. She started illustrating faces on used coffee cups as a means of making art out of something normally thrown away. The designer hopes to continue creating many more.

设计师杰西·吉亚姆布罗尼对咖啡情有独钟，在旧金山设计工作室工作时，她常常喜欢到街对面的咖啡馆中闲坐，享受浓香的咖啡所带来的惬意。她开始尝试在自己使用过的咖啡杯中绘制各种表情，意在赋予那些通常被扔掉的杯子以艺术之美。同时，设计师在未来还会继续坚持这种创作。

Kofi Cult

Design agency: Geraldine Lim
Production date: 2009
Designer: Geraldine Lim
Client: Kofi Cult
Nationality: Australian

Kofi Cult is a Melbourne based take away coffee joint, which believes in quality and that every cup should taste the same. The cup was designed to be eye-catching and iconic, having a fun, friendly and modern twist.

科菲·柯尔特咖啡杯

设计机构：杰拉尔丁·林姆
完成时间：2009年
设计师：杰拉尔丁·林姆
客户：科菲·柯尔特咖啡厅
国家：澳大利亚

科菲·柯尔特是一家坐落在墨尔本的外卖式咖啡厅，该咖啡厅秉承优质的服务理念，坚持确保每一杯咖啡做到无可挑剔。对于这一系列咖啡杯的设计，设计师力图打造一个迷人、醒目、诙谐、友好、时尚的设计方案。

Always Sunny Ice Cream Containers

Production date: 2009
Designer: Danny Maller
Client: Personal work
Nationality: American

The big rule of the project restricted the use of images that were not created by the designer. So he used Illustrator to render his favourite main characters of the show It's Always Sunny In Philadelphia along with memorable quote from the TV series.

"永远阳光灿烂的"冰淇淋容器

完成时间：2009年
设计师：丹尼·玛尔乐
客户：个人作品
国家：美国

该项目的一个重要设计规则限制了一些图片的应用，而这些图片并非出自设计师之手。因此，设计师丹尼·玛尔乐取材自美国情景喜剧"费城永远阳光灿烂"以及一些令人难忘的电视连续剧，巧妙地使用矢量图形处理软件对其喜欢的主要人物角色进行渲染。

Cups & More Cups

Design agency: Paprika Design LTDA
Production date: 2010
Designer: Murilo Kleine
Client: Internal project
Nationality: Brazilian

The collection "Cups & More Cups", developed in 2010, promised the reduction of the waste caused by paper coffee cups within the Paprika Design company. Initially intended as an internal project, the first pictures were iconic characters, hand-drawn on both new and recycled cups.
Based on ecological concepts and low cost production politics, the monochromatic prints also managed to achieve a distinct look and originality.

"杯子&更多的杯子"

设计机构：巴西红辣椒设计有限公司
完成时间：2010年
设计师：穆里罗·克莱
客户：内部项目
国家：巴西

"杯子&更多的杯子"这一项目开发于2010年，旨在减少巴西红辣椒设计有限公司内部的咖啡纸杯浪费问题。该项目起初被视为一个内部项目，首选的图片是一些极具代表性的人物，全部采用手工绘制，材料为全新或回收型纸杯。
设计师们以生态理念和低成本加工费用为出发点，同时运用单色版画成功地实现了纸杯外观的独特性与独创性。

The StarBucks

Production date: 2011
Designer: Kareem Gouda
Client: StarBucks Coffee
Nationality: Egyptian

The designer redesigned the StarBucks Coffee® cups for the middle east region as a side project. the designer created a micro collection of inky coffee cups with black illustrations.

星巴克咖啡杯

完成时间：2011年
设计师：卡里姆·高达
客户：星巴克咖啡厅
国家：埃及

设计师卡里姆·高达为星巴克咖啡厅中东分公司的咖啡厅重新设计了一套咖啡杯，作为一个分支项目，设计师精心创建了一套绘有黑色插画的墨水咖啡杯。

Eleanor Rigbey   "艾莲娜·瑞比"

Production date: 2011   完成时间：2011年
Designer: Klaus Voorman   设计师：克劳斯·沃尔曼
Client: StarBucks Coffee   客户：星巴克咖啡厅

The designer was sitting in utter boredom, flipping through the iPod and sulking over a newly empty Starbucks cup. Suddenly stopping on the classic The Beatles "Eleanor Rigby" he stopped and stared at the coffeeless, pictureless paper cup sitting on the desk. And the idea was born. Using nothing but sharpies, the Beatles Revolver album cover was plastered on the cup before second period.

设计师一边无聊地玩转iPod（苹果公司音乐播放器），一边为空空的星巴克纸杯而恼火时，披头士乐队一首经典的"艾莲娜·瑞比"不禁令他心头一阵，转眼再看办公桌上的咖啡和散落的图片，一个创意理念由此而生。设计师在设计的第二阶段到来之前，巧妙地运用披头士乐队《左轮手枪》专辑的封面，并将其粘贴到纸杯之上。

Shy Octopus   害羞的章鱼

Production date: 2011   完成时间：2011年
Designer: Tara Dayne Walbridge   设计师：塔拉·戴恩·瓦尔布里奇
Client: StarBucks Coffee   客户：星巴克咖啡厅

The designer loves how curious the octopus is. They can both crawl and swim and cling. Some can have spots, or are camouflaged to look like ocean rocks and allege, and some are pink. The designer used a combination of water colour and copic markers to display the differences.

设计师塔拉·戴恩·瓦尔布里奇对聪明而古怪的章鱼情有独钟。它们既会爬行，又能够潜水和黏附。它们的皮肤可以由斑点覆盖，也可以伪装成海洋中的岩石，甚至有一些呈现出粉红色。在此，设计师塔拉·戴恩·瓦尔布里奇巧妙地将水彩和COPIC马克笔相结合，以展现章鱼的变化多端。

Cherry Trees 樱花树

Production date: 2011
Designer: Tara Dayne Walbridge
Client: StarBucks Coffee

完成时间：2011年
设计师：塔拉·戴恩·瓦尔布里奇
客户：星巴克咖啡厅

The designer loves Japanese culture more than any others. The designer loves the flat rice fields, the reaching mountains, the beautiful traditional stone gardens, and of course, the cherry trees. Cherry trees have long been a recognised part of the Japanese culture, every spring a festival is held in their beauty and honour.

设计师塔拉·戴恩·瓦尔布里奇非常热衷于日本文化。他喜欢那里平坦的稻田、高耸的山峰、美妙的传统石花园，当然也包括迷人的樱花树。长久以来，樱花树早已被公认为日本文化的一部分，每年的春天，日本都会举办一个盛大的节日对其进行歌颂。

Waves 波浪

Production date: 2011
Designer: Tara Dayne Walbridge
Client: StarBucks Coffee

完成时间：2011年
设计师：塔拉·戴恩·瓦尔布里奇
客户：星巴克咖啡厅

The designer loves how blues and greens blend in the ocean waves. The water has always caught his attention. It is incredibly hard to draw realistically because of the way the light bends inside of it. It can range anywhere from light blue to emerald to deep indigo, so dark that you can barley see. The designer used watercolour paint to show a soft texture, just as the water itself is soft.

设计师塔拉·戴恩·瓦尔布里奇对海浪蓝、绿色相交的场景深深迷恋。水面常常是其所关注的焦点。在现实中，将这一美妙的场景进行描绘是非常困难的，因为常常受到光线弯曲方式的制约。光线弯曲的范围涉及浅蓝、翠绿色以及深靛蓝色等。在此，设计师巧妙运用水彩画以展现一个柔软的质地，从而使其与柔软的水体相得益彰。

Starry Night

Production date: 2011
Designer: Tara Dayne Walbridge
Client: StarBucks Coffee

Van Gogh's beautifully abstract Starry Night is an instant classic in art, and a personal favourite of the designer. He fell in love with this amazing piece of work after painting an imitation for the first time. He doesn't think he could ever appreciate a work of art more than he does this one. The designer painted this piece out of respect of it and its artist. The picture is painted in watercolour for layering purposes.

星空

完成时间：2011年
设计师：塔拉·戴恩·瓦尔布里奇
客户：星巴克咖啡厅

艺术大师文森特·梵高的精美抽象画作"星空"堪称是艺术届的一个经典之作，同时也深受设计师的喜爱。设计师在首次模仿这一画作之后即被这一迷人的画作所深深感染。在他看来，从没有一幅这样的艺术画作能够让他如此痴迷。对于这幅画作的模仿完全出自其对这一画作及其创作者的尊敬。这幅水彩画的绘制以分层为目的。

Romeu & Julieta

Production date: 2011
Designer: Adriana Amaral Pepplow, Valdir de Oliveira
Client: Kibon
Nationality: Brazilian

The project is a conceptual package design for a fictitious ice cream. The idea was based on creating an alternative of sweet gift for Valentine's Day. Romeu & Julieta is not only the Shakespeare's classic novel, but in Brazil it is also a traditional flavour of dessert made with white cheese and guava paste. To mix the ideas of flavour and romance, this package features the job of a couple of designers, in which Adriana Amaral was responsible for the art direction and Valdir de Oliveira was responsible for character design and 3D simulation.

罗密欧与朱丽叶

完成时间：2011年
设计师：阿德里安娜·阿马拉尔·佩普罗，瓦尔迪尔·德·奥利韦拉
客户：奇本冰淇淋公司
国家：巴西

该项目是一个虚构的冰淇淋包装设计概念。创意理念是以创建一个独一无二的情人节甜美礼物为基础。罗密欧与朱丽叶不仅是莎士比亚经典小说中的人物，在巴西，也是一种以白奶酪和番石榴酱为原料的传统风味甜点。为了将独特的口味与浪漫的含义相结合，这一包装方案特别安排两位设计师共同合力完成，其中，设计师阿德里安娜·阿马拉尔·佩普罗负责艺术指导工作，而瓦尔迪尔·德·奥利韦拉则主要进行人物设计和三维仿真工作。

Senses Awake

Design agency: Quantum Graphics
Production date: 2011
Designer: Irina Kryucheva
Client: Interpack 2011 for PDA Innovation booth
Nationality: Russian

Awakening of senses is the main motif behind the concept. It tells a story of the beautiful girl character with the help of an interactive hook: when the package is closed the girl is "asleep" and her eyes are closed. But as soon as you open the package, the girl "wakes up", attracted by the wonderful coffee aroma. The main "hero" of the dispenser for coffee pads is an inspired coffee cup with wings. The cup is placed around the die-cut window making play of the contents and of the form.

感官的苏醒

设计机构：量子图形设计工作室
完成时间：2011年
设计师：伊莉娜·科瑞彻娃
客户：2011年英特帕克PDA创新展位
国家：俄罗斯

感官的苏醒是这一设计概念背后的主题。这款包装在交互挂扣的帮助下，讲述了一个关于美丽的女孩的故事：当包装盒呈闭合状态时，女孩就"睡着了"，并且眼睛是闭着的。但是只要你打开包装盒，女孩就"醒来了"，她是被美妙的咖啡香气所吸引。咖啡杯垫上"调剂师仙"的设计灵感来源于带翅膀的咖啡杯。将杯子置于打孔窗子周围以便凸显杯子的外形设计和容量。

Ground Coffee Amado

Design agency: Quantum Graphics
Production date: 2010
Designer: Irina Kryucheva
Client: CoffeeArt
Nationality: Russian

In Amado is the real coffee professionals and Amado want consumers to be aware. They are happy to share with them the deep knowledge about coffee. With the help of the package Amado demonstrate to consumers all basic stages of coffee production. Each of 7 SKU's has a picture of one of the main coffee production stages: growing it up, collecting, drying, frying, etc. Big bright pictures, headlines, text blocks create the look of travel magazines, very popular amongst Amado target.

阿玛多研磨咖啡

设计机构：量子图形设计工作室
完成时间：2010年
设计师：伊莉娜·科瑞彻娃
客户：咖啡艺术
国家：俄罗斯

阿玛多咖啡是真正的专业咖啡，并且阿玛多咖啡希望消费者能够掌握关于咖啡的知识。他们非常高兴同大家分享关于咖啡的更多知识。在该包装设计的帮助下，阿玛多咖啡向消费者展示了所有的咖啡生产的基本阶段。每一款7 SKU's咖啡上都有着一张展示着主要的咖啡生产阶段的插图：种植、收集、干燥、油炸等。色彩明丽的大篇幅照片、头条标题和文本块的设计给人们带来一种阅读旅游杂志的感觉。

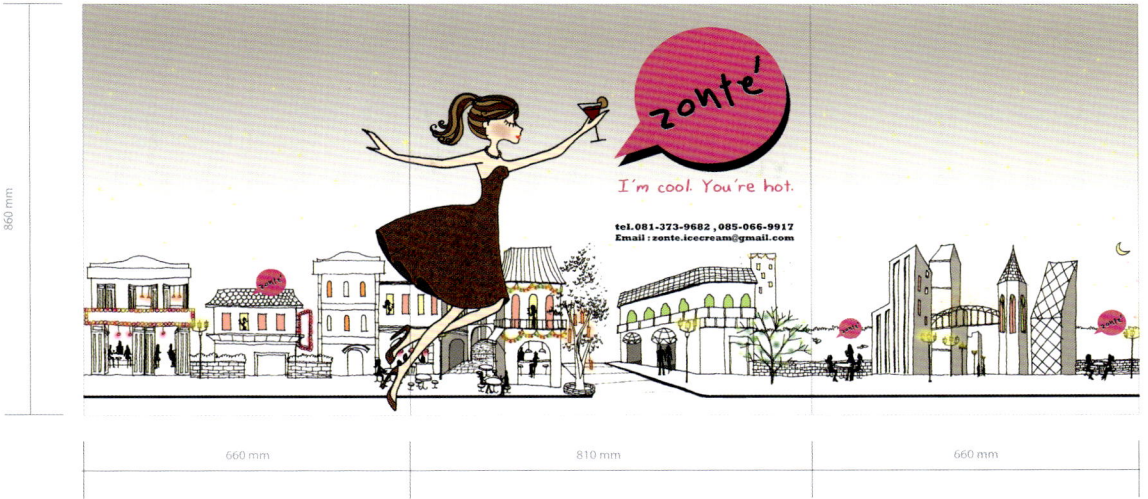

Zonte'

Production date: 2008
Designer: Piraya Ruangpungtong
Client: Sipim Wanwieng
Nationality: Thai

Zonte' is a brand of cocktail ice cream from small entrepreneur which will distribute at restaurants in Bangkok. The client required a mood of night-life-city in France to represent the brand, and happy sexy girl as a mascot. The designer differentiated the design by using hand illustration applying to logo and the promotional design such as menu, freezer and ice cream paper cup. All the ice cream flavours use the same design cup, classified by the tick-box on the cover.

Zonte'冰淇淋

完成时间：2008年
设计师：皮拉娅–阮彭东
客户：斯皮姆·万翁
国家：泰国

Zonte'是一个小型鸡尾酒冰淇淋品牌，其顾客主要面向曼谷的餐馆。客户委托设计师为其设计一款能够代表该品牌的法式城市夜生活设计方案，并塑造出一个性感女郎作为象征物。因此，设计师运用手绘的方式，对应用到标识和宣传设计材料（菜单、冰柜以及冰淇淋纸杯）上的设计进行区分。所有的冰淇淋口味均使用相同的设计纸杯，仅使用封顶上的标记进行区分。

An Image

Production date: 2008
Designer: Tony Stavrou
Client: Personal work
Nationality: British

When the designer was in a fast food restaurant, after having lunch, watching the store's posters, he drew on the cup. Then back to his studio, the designer shoot it down, uploaded the iamge to computer and made corrections with photoshop. He put it on the desk, and each time when he saw it he would recall the fast food restaurant. Random words and simple graphics give a relaxed feeling.

"一个图像"

完成时间：2008年
设计师：托尼·斯塔瓦罗
客户：个人作品
国家：英国

有一次，设计师托尼·斯塔瓦罗在一个快餐店就餐，在用完午餐之后，仔细端详店内的海报，顺便在纸杯上涂鸦。当他返回设计工作室之后，细心地将这一设计拍摄下来，随后将其下载到电脑之上，并运用图形处理软件（photoshop）进行修正。创作完成之后，他将其放在了办公桌上，每次看到这一设计都会让他想起那家快餐店。随意的语言、简单的图形传递出一种轻松之感。

Flour Coffee Paper Cup

Design agency: A-W Design
Production date: 2011
Designer: Leona Lewis
Nationality: British

The coffee cup is designed for Flour Coffee Shop. The coffee shop owner likes this pattern, so the designer used it in the coffee shop brand design. To complement the cup design, the designer chose black for the background colour to highlight the pattern.

面粉咖啡店

设计机构：A-W设计工作室
完成时间：2011年
设计师：丽安娜·刘易斯
国家：英国

这款咖啡杯是为了面粉咖啡店而设计的。咖啡店的老板很喜欢这个图案，因此设计师将这个图案用在了咖啡店的品牌设计上。作为对整个设计的补充，设计师最终选择了黑色，以凸显整个图案。

Greenpaper Cup

Production date: 2009
Designer: Adam Michael
Client: Personal work
Nationality: Sweden

Coffee cup in our lives plays an important role. But most of us throw them away just after the finish. Adam Michael designed this series of green cups, just to let everyone keep the cups after drinking. Pale green, overlapping leaves want people to use them with a good mood. Now we are advocating green. The designer hopes the users won't casually discard the cups after use.

绿色纸杯

完成时间：2009年
设计师：亚当·迈克尔
客户：个人作品
国家：瑞典

咖啡杯在我们的日常生活中扮演了一个重要角色。然而，大多数人在对其使用之后不会投入太多的关注。设计师亚当·迈克尔设计的这一系列绿色纸杯，意在促使大家在品尝咖啡的同时对咖啡杯的设计细心观察。浅绿色的色调搭配重叠的叶片将为使用者带来一个美好的心境，同时与环保的绿色理念相得益彰。设计师希望这一精心设计的纸杯将促使使用者们不再将其随意丢弃。

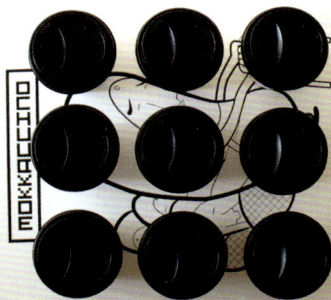

Mokkaccino Crafts

Design agency: HiredMonkeez
Production date: 2009
Designer: Edmundo Moi-Thuk-Shung
Client: Personal work
Nationality: British

The characters on the art cover have a background story for a game the designer will be working on in the future. It's about coffee beans which have magic powers, the "Mokkaccino beans". The cups were steady and the prints had "okay" quality. It also stayed on when you'd poured hot water in the cups.

摩卡奇诺工艺咖啡杯

设计机构：HiredMonkeez
完成时间：2009年
设计师：埃德蒙多·莫伊–图克–崇
客户：个人作品
国家：英国

这一艺术封面上的人物拥有一个有趣的游戏背景故事，而这一游戏是设计师未来所从事的工作之一。故事讲述的是一群拥有神奇魔力的咖啡豆，它们拥有一个美妙的名字 —— "摩卡奇诺家族"。这一系列纸杯外形稳固，拥有优秀的印刷品质。当杯中盛满温热的液体时，这些印刷体也依然显现。

Frozen Yogurt

Design agency: OBAH Design
Production date: 2011
Designer: Luciano Ferreira
Client: Study project
Nationality: Brazilian

This study presents proposals for a line of frozen yogurt. Based on the concept of local crafts of Minas Gerais (BR), the aesthetic structure was developed with elements of the culture of this Brazilian state: quilts, gossip, and much of the local craft served as a symbolic reference to the design.

冷冻酸奶

设计机构：OBAH设计工作室
完成时间：2011年
设计师：卢西亚诺·费雷拉
客户：研究项目
国家：巴西

这一研究项目展现了设计师卢西亚诺·费雷拉对系列酸奶产品的设计方案。设计以当地的米纳斯·吉拉斯州（巴西）的工艺品为设计基础，此外，美学构架的开发运用了巴西国家的文化元素，即拼布床单，闲谈，而大多数的地方工艺品还被引入到设计之中作为一个象征性的参考。

Stories

Design agency: BVD
Production date: 2007
Client: Turesgruppen AB
Nationality: Sweden

The design is to create a strong and totally unique café experience: from concept and name, to graphic profile and packaging. The concept needed to be warm, welcoming, honest and genuine and targeted to young professionals.

Black, white and stainless steel is blended with warm wood, and the old fashioned café-feeling expressed by things like a board with old, detachable letters and traditional cups and trays. The graphics are clean and simple, but at the same time surprising and playful. The design exudes personality, quality, style and a big city feeling.

故事

设计机构：BVD设计工作室
完成时间：2007年
客户：Turesgruppen AB咖啡厅
国家：瑞典

该项目的设计目标是创建一个强大而独特的咖啡体验：包括品牌理念和名称、平面描述和包装。设计理念要求传递出温馨、热情、真诚、纯正之感，以年轻的上班族为主要销售对象。

黑色、白色色调与不锈钢材料以及温暖的木料完美结合，带有复古、可拆分字母的黑板、传统的咖啡杯和托盘等悄悄地为整个空间营造出一种老式咖啡馆的空间氛围。图形的设计清晰而简洁，同时不乏惊喜与诙谐之感。

EN SÅDAN HÄR MUGG MED KAFFE RÄDDADE MAGNUS   MAGNUS FRÅN ATT SOMNA.

STO ORIE ES S

Repurpose Insulated Cup

Production date: 2011
Designer: Vera Valentine
Client: Repurpose Compostables
Nationality: American

Repurpose compostables enlisted designer Vera Valentine to design their world's greenest coffee cup which required no sleeve or double cupping and was made entirely from plants.

"可堆肥再利用"公司的隔热杯

完成时间：2011年
设计师：维拉·瓦伦汀
客户："可堆肥再利用"公司
国家：美国

来自美国的"可堆肥再利用"公司委托设计师维拉·瓦伦汀为他们提供一款全球最具环保意识的咖啡杯设计方案，另外，还特别注明，该纸杯没有杯套或双层杯壁，完全以植物为原料。

Mother Burger Packaging Take-Away Food

Design agency: Studio Kluif
Production date: 2009
Designer: Jeroen Hoedjes
Client: Mother Burger
Nationality: American

Mother Burger is a N.Y.-based health-conscious fast-food restaurant chain. Their mascot is "Mother", a large biker kind of guy who is committed to serving up wholesome quality food that is of organic origin and without committed or hormones. It is a burger joint with an attitude and a sense of humour. Mother will make sure you get all the 4 food groups: Burger, Hot Dog, Fries and Beer. Pity there aren't more mothers like that, we think.

"母亲汉堡"外带食品包装

设计机构：克鲁夫设计工作室
完成时间：2009年
设计师：吉荣·霍德耶斯
客户：母亲汉堡
国家：美国

"母亲汉堡"是纽约一个弘扬健康饮食的快餐连锁餐厅。该餐厅的吉祥物是一位体形硕大的骑车人承诺为顾客提供有益健康的优质食品，并强调这些食品完全采用有机材料制成，毫无不健康因素或者激素。这是一家汇聚了谦和与幽默感的汉堡店。在这里，顾客将能够购买到四种食品，即汉堡、热狗、薯条和啤酒。很遗憾，这样的"母亲"并不多。

MOTHER SAYS:
MAKE SURE YOU
GET ALL FOUR
FOOD GROUPS:
● BURGER
● HOT DOG
● FRIES
✓ BEER

The Scoop

Design agency: The Creative Circus
Production date: 2011
Designer: Charis Ceniseroz
Client: Breyers Ice Cream
Nationality: American

Friendships are often formed over a common interest. The Scoop, a sub-brand of Breyers, uses specially designed packaging to raise funds for ice cream socials between kids and seniors. These socials create a place where new friendships are formed, and participants can share ice cream, and their stories.

"舀勺"冰淇淋

设计机构：创意广场设计工作室
完成时间：2011年
设计师：查理斯·塞内斯塞洛兹
客户：布雷耶冰淇淋
国家：美国

往往，共同的兴趣爱好能够促进友谊的形成。"舀勺"作为布雷耶冰淇淋的一个旗下品牌，意在利用独特的包装为儿童和老年人冰淇淋联谊会凑集资金。这些联谊会为人们建立友谊搭建了一个理想平台，在此，人们可以在品尝冰淇淋的同时，分享他们的故事。

Grind Espresso Bar

Design agency: A-side Studio
Production date: 2011
Designer: James Booth
Client: Grind Espresso Bar
Nationality: British

The designer was asked to design a branding proposal for a new Espresso Bar in Truro, Cornwall. The design was made as bold and visible as possible, in order for it to compete with existing coffee shops. Take-out paper cups were a great way to promote the bar as they would be able to be seen as they were carried by customers around the city.

研磨咖啡吧

设计机构：A座设计工作室
完成时间：2011年
设计师：詹姆斯·布提
客户：研磨咖啡吧
国家：英国

在英国康沃尔郡特鲁罗市新落成的研磨咖啡吧委托设计师詹姆斯·布提为其设计一款品牌塑造方案。设计的风格要求大胆而醒目，从而与已有的那些同类咖啡馆形成鲜明的对比。由于外带式纸杯是宣传咖啡吧的一个重要途径，它们将被顾客带到任意一个角落，因此，其设计被倾注了设计师极大的热情和精力。

Kernel Kustard Cups

Design agency: Mitre Agency
Production date: 2007
Designer: Ryan McCullah
Client: Kernel Kustard
Nationality: American

Kernel Kustard is as close as you'll get to Chicago eats outside of the windy city. With authentic, rich kustard, Chicago dogs and gourmet popcorn, Kernel Kustard has quickly become a popular local favourite. The packaging combines both the fun and grit of Chicago with bold overlapping "carnival" type, topped off with the friendly and iconic Kernel Kustard logo.

柯乃尔·库斯塔德纸杯

设计机构：米特芮设计公司
完成时间：2007年
设计师：瑞安·麦克库拉
客户：柯乃尔·库斯塔德餐厅
国家：美国

柯乃尔·库斯塔德餐厅将是你在风城芝加哥户外就餐的一个最亲密场所。凭借出售口味纯正而浓厚的糕点、芝加哥热狗以及美食爆米花等食品，柯乃尔·库斯塔德餐厅已迅速发展成为当地最受欢迎的餐厅之一。对于该品牌的包装设计，设计师巧妙地将芝加哥的意趣和沙砾与大胆的重叠式"狂欢节"字样完美结合，并冠之以友善、醒目的柯乃尔·库斯塔德餐厅标识。

ABC Paper Cup

Design agency: He was born
Production date: 2008
Designer: Sunhan Kwon
Client: He was born
Nationality: South Korean

"ABC Paper Cup" takes a break from the conventional paper cup design that sometimes makes you confused with which one you are using when you are having a party or etc. The ABC Paper Cup suggests you how to remember your own cup easily by featuring a unique design of fourteen varieties with English alphabets from A to Z. The promotion clip includes the design of ABC Paper Cup as well as fun suggestions on how to use ABC Paper Cups. 30 paper cups are randomly mixed in a box regardless of alphabetical order. It was awarded the winner in the red dot design award 2009.

ABC纸杯

设计机构："He was born"设计工作室
完成时间：2008年
设计师：权尚韩
客户："He was born"设计工作室
国家：韩国

传统的纸杯在聚会或其他场合时经常容易被拿错，从而为使用者带来困扰，而"ABC纸杯"突破了这一格式化设计，通过巧妙运用英文字母A到Z的十四个变体，使每个纸杯都拥有各自的特色，从而不再轻易地被拿错。除此之外，设计师还专门提供了宣传短片的设计，其中包括ABC纸杯的设计以及使用ABC纸杯的有趣建议。30个纸杯不按字母顺序随意地摆放在一个盒子中。该项目曾在2009年荣获了红点设计大奖。

Batch® Ice Cream

Design agency: Devers Ink&Lead
Production date: 2010
Designer: Wade Devers
Client: Batch® Ice Cream
Nationality: American

Batch is made by hand in small quantities in Jamaica Plain, Massachusetts in the United States. The product is made according to a very simple philosophy "good food should be good for you". Each container of this artisan-quality ice cream is made using only the freshest local ingredients. Drawing influence from city neighbourhood restaurants, which write their menus each day on chalkboards, the designers were able to reflect the handcrafted, local and fresh qualities of the product. The paper cup packaging provides a tactile and familiar experience, reminding people of the ice cream treats of their youth.

贝奇®冰淇淋

设计机构：德弗斯油墨&铅设计工作室
完成时间：2010年
设计师：韦德·德弗斯
客户：贝奇®冰淇淋
国家：美国

贝奇是美国马萨诸塞州牙买加平原上一个小型手工制作冰淇淋店。店内加工并出售的食品以一个简单的理念为基础，即"优质的食物有益于身体的健康"。每一款精致冰淇淋容器的打造均取自当地的新鲜材料。设计师在对城市周边餐厅进行考察后发现，大多数餐厅喜欢将他们每天的菜单写在黑板上，因此，设计师巧妙地将产品的手工制作特色、地方风格以及新鲜的品质完全呈现在包装容器之上。匠心独运的纸杯包装将为顾客营造一个亲切而熟悉的感知体验，唤起人们对童年冰淇淋味道的回忆。

House of Arts Berlin

Design agency: wernery.com
Production date: 2010
Designer: Benjamin Wernery
Client: Bachelor project
Nationality: Danish

Product design of a corporate identity project (House of Arts Berlin — Bachelor). "House of Arts Berlin" is a gallery house of experience, which opens itself up for collectors, curators, and anyone exploring the broad field of art.

柏林艺术之家

设计机构：wernery.com设计工作室
完成时间：2010年
设计师：本杰明·维纳里
客户：学士项目
国家：丹麦

这是一个企业识别项目（柏林艺术之家——学士）中的一个产品设计方案。"柏林艺术之家"是一个体验画廊空间，主要面向艺术领域的收藏家、策展人、探索者开放。

Candy King

Design agency: BVD
Production date: 2005
Designer: Bengt Anderung
Client: Candyking International Ltd
Nationality: Sweden

BVD created the "Candy King's kingdom", from which all communication would emanate. The king himself was given a more contemporary appearance and a friendlier personality. The colours communicate joy and variety.

"糖果国王"

设计机构：BVD设计工作室
完成时间：2005年
设计师：本特·安德郎
客户：糖果国王国际有限公司
国家：瑞典

这一项目是BVD设计工作室专为糖果国王国际有限公司而提供的品牌塑造方案，包括所有宣传材料的设计。其中，"国王"被赋予了一个较为时尚的形象和一个友善的个性。活泼的色调完美地传达出愉悦与丰富之感。

froyo

Design agency: asprimera design studio
Production date: 2005
Designer: Eleonora Xanthopoulou
Client: froyo
Nationality: Greece

Froyo is the first store of frozen yogurt in Athens, Greece. Paper cups were designed to communicate fresh and healthy product in a modern way. The designer includes minimal typography and design based on the spoon that someone uses to enjoy froyo.

"冰冻酸奶"

设计机构：asprimera设计工作室
完成时间：2005年
设计师：埃莱奥诺拉·桑多普罗
客户：冰冻酸奶
国家：希腊

"冰冻酸奶"是希腊雅典的首家冰冻酸奶经营店。对于该店纸杯的设计，设计师力图运用一个时尚、现代的方式传达出该品牌产品的新鲜与健康特色。小写的字体和颇具设计感的舀勺为享用"冰冻酸奶"的顾客营造出喜悦之感。

Honey & Mackie's

Design agency: Wink
Production date: 2011
Designer: Scott Thares
Client: Honey & Mackie's
Nationality: American

Honey & Mackie's is an ice cream shop for kids that caters to parents. The name of the establishment comes from the nicknames of the owner's children. Thus, the branding and packaging needed to be modern, authentic and kid fun. The ingredients are all natural, organic and locally grown.

汉尼&摩奇冰淇淋店

设计机构：闪烁设计工作室
完成时间：2011年
设计师：斯科特·泰瑞斯
客户：汉尼&摩奇冰淇淋店
国家：美国

汉尼&摩奇是一家儿童冰淇淋店，主要面向有宝宝的爸爸妈妈们。这一店名的确立取材自店主孩子们的乳名。因此，对于品牌塑造与包装方案的设计，设计师力图使其充满时尚、单纯而童真之感。在此出售的冰淇淋产品均取自纯天然、有机、源自当地的材料。

Nonna Lina Gelato Italiano

Design agency: Mr. Conde Studio
Production date: 2011
Designer: Rafael Conde
Client: Nonna Lina Gelato Italiano
Nationality: Brazilian

Mr. Conde Studio created and developed a special cup design for ice cream brand Nonna Lina, a homemade Italian ice cream brand that has developed its revenues more than 60 years. This tradition was the starting point for creating the brand and its packaging. Made with the finest natural ingredients and grout developed in southern Italy, Nonna Lina ice cream has a philosophy to be the best of the world with the best flavour, better texture and promises to offer a passion for candy ice cream.

诺娜·丽娜意大利冰淇淋

设计机构：康德先生设计工作室
完成时间：2011年
设计师：拉斐尔·康德
客户：诺娜·丽娜意大利冰淇淋
国家：巴西

康德先生设计工作室专为诺娜·丽娜意大利冰淇淋设计并开发了特殊包装的纸杯方案。诺娜·丽娜是一个自制的意大利冰淇淋品牌，至今已拥有60多年的创办历史，而这一悠久的传统恰恰也成为该品牌及其包装设计的出发点。取材自优质天然成分和南部意大利地区原浆的诺娜·丽娜意大利冰淇淋以经营全球口味最佳、口感上乘的冰淇淋为经营理念，承诺对每个糖果冰淇淋的加工投入全部的热情，并将这种热情一直延续到顾客的手中。

Cielito Querido Café

Design agency: Cadena + Asociados Branding
Production date: 2010
Designer: Rocio Serna
Client: Cielito Querido Café
Nationality: Mexican

Cielito® is a Latin American reinvention of the coffeehouse experience. It is a place that surprises, conforts and engages all senses through its space, aroma, taste, colour, and histories. The design draws its inspiration from Mexican history: the games, the joyful colours, the language of symbolism, and the illustrated graphics of the late 19th to early 20th century.

亲爱的谢利托咖啡厅

设计机构：卡德纳品牌设计联合公司
完成时间：2010年
设计师：罗西奥·塞尔纳
客户：亲爱的谢利托咖啡厅
国家：墨西哥

谢利托咖啡厅是一个拉美风格的咖啡厅改造项目。该空间中充满着惊喜、舒适的味道，凭借独特的空间、香气、口感、色泽和历史为顾客提供了一个独特的感官体验。这一系列纸杯的设计灵感源自设计师对墨西哥历史的参考，涉及游戏、快乐的色彩、象征性语言以及19世纪末至20世纪初的插画图案等。

Caffeine isn't a drug, it's a vitamin!

LETHAL!

SOS! Coffee

Production date: 2011
Designer: Cat Dempsey
Client: SOS! Coffee
Nationality: Irish

The aim of this project was to create an identity and packaging for an Irish coffee shop chain called SOS! Coffee. The name came from the Irish word "SOS" which means break, so the name translates as coffee break. It also expresses the urgency with which people need their coffee.

SOS!咖啡厅

完成时间：2011年
设计师：凯特·登普西
客户：SOS！咖啡厅
国家：爱尔兰

这一项目的设计目的是为爱尔兰SOS！连锁咖啡厅设计一个识别与包装方案。"SOS"作为该餐厅的名称在爱尔兰语中含有"休息"之意，因此，这一名称也可以被翻译成"休息时间"。与此同时，这一名称也表达了顾客对咖啡的渴求。

Vida e Caffé

Design agency: Studio Muti
Production date: 2011
Designer: Clinton Campbell
Client: Vida e Caffé
Nationality: South African

The brief was to design a sleeve to celebrate Vida e Caffé's 10th anniversary. The designers used phrases you would encounter in a coffee shop, a mixture of Portuguese and African languages. They used a hand-written style and a misprinted effect to stay true to the African ruggedness.

维达咖啡厅

设计机构：穆蒂设计工作室
完成时间：2011年
设计师：克林顿·坎贝尔
客户：维达咖啡厅
国家：南非

这一项目的设计理念是为庆祝维达咖啡厅成立十周年而设计的一款杯套。来自穆蒂设计工作室的设计师们巧妙运用这一咖啡店中常用的葡萄牙和非洲语言的合成短语。整个设计采用手写的风格，与印错的视觉效果一同彰显出非洲人粗犷、强硬的作风。

Toni's Ewe's and Goat's Milk Yoghurt

托尼家母羊与山羊酸奶

Design agency: moodley brand identity
Production date: 2011
Designer: Natascha Triebl
Client: Toni's Handels GmbH
Nationality: Austrian

设计机构：莫德利品牌标识设计工作室
完成时间：2011年
设计师：娜塔莎·特里布尔
客户：托尼家交易公司
国家：奥地利

The project is the cup design for Toni's Ewe's and Goat's Milk Yoghurt. Products of the brand Toni's have always been standing for the theme of "freedom" – since the brand is the pioneer of the famous organic free-range eggs in Austria. Also Toni's and goats enjoy their life under the heavens. Typography, illustrations and colours of the new 100% recyclable packaging for ewe's and goat's milk yoghurt have high brand recognition and are also connected with the theme of freedom. Other issues such as health, nature and animal protection are conveyed in a beastly funny way.

这一项目是莫德利品牌标识设计工作室专为托尼家母羊与山羊酸奶品牌而打造的纸杯设计方案。长期以来，由于托尼品牌是奥地利著名的家禽自由放养并加工有机鸡蛋的先锋，因此，一直被视为"自由"主题的代表。同时，托尼强调其原材料的来源必须纯天然。因此，这一系列包装设计完全采用可回收材料，搭配独特的字体、插画和色调，在突出品牌识别性的同时，与该品牌所弘扬的自由主题相得益彰。除此之外，健康、天然、保护动物等话题也巧妙地以一种拟人的方式被传达出来。

Instant Sidekick

Production date: 2010
Designer: Crisy Meschieri
Client: Personal work
Nationality: Argentinean

Even heroes need a sidekick. Intended for our everyday heroes (firefighters, nurses, doctors), Instant Sidekick is an extra-caffeinated coffee that can be instantly called upon to help save the day. The disposable cups have packets of coffee, creamer, and sugar attached to the bottom so that all the user has to do is pour in hot water, pop off the stirrer from the top of the cup, and go!

"速溶密友"

完成时间：2010年
设计师：克里塞·梅斯彻理
客户：个人作品
国家：阿根廷

即使是英雄，也需要一个密友。专为生活中的英雄（消防员、护士、医生）而服务的"速溶密友"是一个口感极佳的咖啡品牌，能够及时满足顾客的需要，从而帮助他们度过美好的一天。这一系列一次性纸杯中在杯底放置了一个咖啡包、奶精和蔗糖，使用者只需向其中倾倒热水，用搅拌器从杯子上方进行搅拌后即可饮用！

Coffee Packaging

Design agency: Ieuw Design
Production date: 2011
Designer: Ilse van der Velde
Client: Personal work
Nationality: Dutch

The new coffee packaging is designed to appeal to younger people and make them see that drinking coffee is not just for your parents. It has a more modern and hip look, both in graphic design as in packaging. It is a hip coffee mug.

咖啡包装

设计机构：Ieuw设计工作室
完成时间：2011年
设计师：埃尔赛·凡·德·维尔德
客户：个人作品
国家：荷兰

这一新式咖啡包装方案旨在吸引青年群体的目光，向他们证明喝咖啡并不仅仅是他们父母的权力。这一系列纸杯的图案和包装设计均完美地流露出现代而时尚的气息。无疑这是一款时尚的咖啡杯！

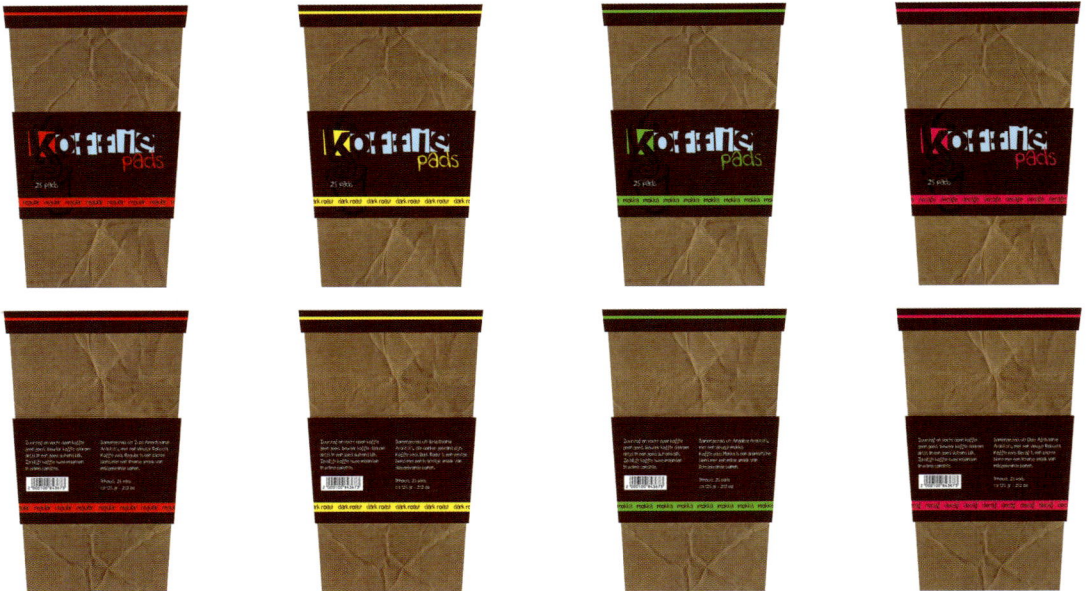

Union Yard

Design agency: The Click Design Consultants
Production date: 2012
Designer: Matt Hancock
Client: Union Yard
Nationality: British

Union Yard is a new addition to the thriving café and deli community of Norwich. Keen to stand out from the outset, Union Yard approached The Click to create a brand under which the store could flourish. The ubiquitous corrugated card cup served as inspiration for the Union Yard logo. This bold, simple, versatile form is applied across the full range of products sold in store, including artisan pastries, fresh sandwiches and locally sourced produce. Outside, exterior signage gives Union Yard a strong presence on a busy intersection while in store, customers are kept informed of what's on offer via clear and concise menu boards.

Union Yard熟食店设计

设计机构：咔哒设计顾问公司
完成时间：2012年
设计师：马特·汉考克
客户：Union Yard熟食店
国家：英国

Union Yard是一家新兴起的位于英国诺威奇市的小餐馆与熟食店社区。热衷于与众不同的Union Yard找到了咔哒设计顾问公司来创建一个能够使商店蓬勃发展的品牌形象。将普遍使用的瓦楞纸杯作为Union Yard熟食店标志的设计灵感来源。这种大胆的、简单的、通用的形式被应用在出售的整个系列的产品上，包括工艺糕点、新鲜的三明治以及就地取材制成的产品。在商店外，标牌为处在繁忙十字路口的熟食店赋予了一种强大的存在感。而在店里，顾客通过清晰、简明的菜单板就能够获取到商店的产品信息。

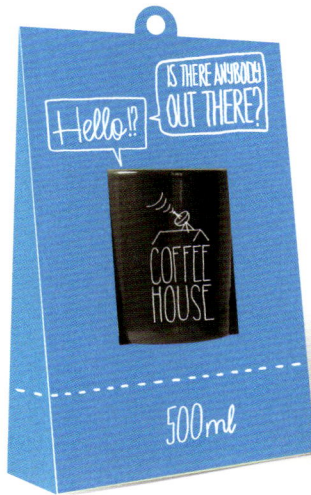

Coffee House

Production date: 2011
Designer: Dmitry Vasiliev
Client: Personal work
Nationality: Russian

Coffee House is an accessible and friendly place. It is for those people who share optimistic and self-ironic views. For example, morning rush, is the great time for self irony, and the cup of coffee can be the part of this.

咖啡屋

完成时间：2011年
设计师：德米特里·瓦西里耶夫
客户：个人作品
国家：俄罗斯

"咖啡屋"是一个亲切、友好的咖啡空间。此处是人们分享快乐与自我解嘲的一个理想场所。例如，早晨上班高峰一直是自我嘲讽的一个谈资，而此时此刻，咖啡必不可少。

eighthirty

Design agency: noahbutcher
Production date: 2010
Client: eighthirty
Nationality: New Zealand

In an attempt to create "stand out" in a cluttered market, the inspiration came from way outside of the category; medicine bottles and even dry cement packaging were being tossed around. This reference inspired eighthirty™ to have a simple colour palate, while the ephemeral dialogue and typographic treatment allow each pack to be unique. Without bowing to the normal conventions of the "Fair trade organic" aesthetic, the packaging attempts to subvert this norm by being clean and simple, with a little bit of cheekiness for good measure.

eighthirty咖啡杯

设计机构：noahbutcher设计工作室
完成时间：2010年
客户：eighthirty咖啡
国家：新西兰

该项目的设计目的是创建一个独一无二的包装方案，从而使其在拥挤的同类商品市场上独领风骚。这一项目的设计灵感源自这一范畴以外的行业；药瓶乃至干式胶水包装容器随处可见。鉴于此，设计师选用了一个简约的配色方案，而一次性的对话和字体处理模式更加确保了每个包装的独特性。这一包装方案并没有遵循传统"正常公平交易"的惯例，而是试图以一个干练、简约的形象对其进行颠覆，同时不乏大胆、前卫的气势。

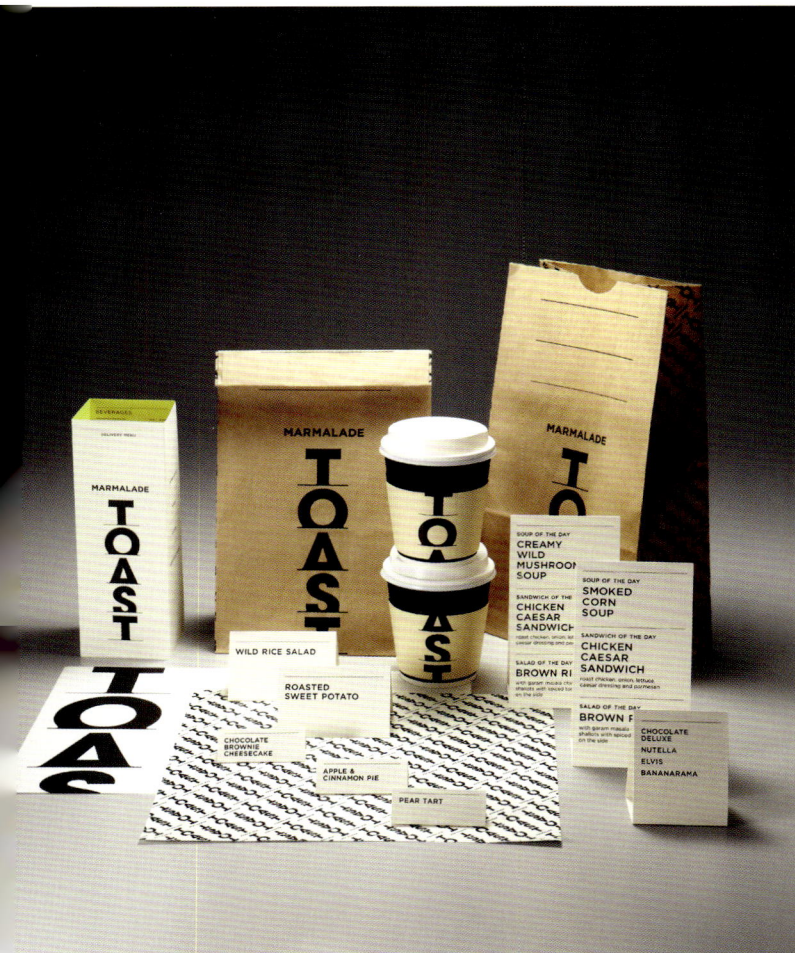

Marmalade Toast

Design agency: &Larry
Production date: 2010
Designer: Lee Weicong
Client: Marmalade Toast
Nationality: Singaporean

This is a fresh new identity for an upmarket gourmet café from The Marmalade Group. Previously known as "Toast", the café's brandmark has been refreshed to include "Marmalade" as a headline and co-branding element. The letters for "TOAST" are rendered vertically in a custom typeface with truncated baselines, reminiscent of bread slices popping out of a toaster. Slightly rounded-off corners at the bottom of the letters are reminiscent of melted cheese on toast.

果酱吐司咖啡厅

设计机构：&Larry设计工作室
完成时间：2010年
设计师：李伟聪
客户：果酱吐司咖啡厅
国家：新加坡

该项目是设计师专为"果酱集团"旗下一个高级美食咖啡厅而设计的一个全新识别方案。这一咖啡厅的原名为"吐司"，其品牌商标经改造后被添加了一个"果酱"字样，并将其作为一个联合品牌元素。英文"TOAST"（果酱）的每一个字母均被赋予了一个带有截断基线的特制字体，并以垂直的形式进行排列，这一精心设计令人自然联想到从面包机中弹出的面包片。与此同时，字母底端略带弧度的转角也令人不禁联想起吐司上融化的奶酪。

Sorvete Premium

Design agency: OBAH Design
Production date: 2011
Designer: Luciano Ferreira
Client: Supermercado Verdemar
Nationality: Brazilian

The design was made for a supermarket, Verdemar that wanted to create a Premium category of their ice cream. The challenge was to create an identity that could feel exclusive, homemade and reflect the idea of their new gourmet flavours. The OBAH Design used a retro aesthetic within a contemporary style. For the product to outstand in the Brazilian market the designers used colours only as patterns, a different one for each flavour. The typography and the illustrations were also made for this clean concept that helped the package be unique.

优质冰淇淋

设计机构：OBAH设计工作室
完成时间：2011年
设计师：卢西亚诺·费雷拉
客户：海绿色超级市场
国家：巴西

这一项目是OBAH设计工作室专为海绿色超级市场旗下一个优质冰淇淋系列产品而设计的包装方案。设计的挑战是创建一个独特的标识，既能够突出产品的独一无二以及手工制作的特点，又能够确保将这一品牌全新的美食风味理念进行有效传达。来自OBAH设计工作室的设计师在一个时尚的格调中运用了一个复古的美学理念。为了使这一系列产品能够在巴西市场上独领风骚，设计师们仅仅运用色彩来代替图案，并确保每种口味都拥有一个不同的包装风格。除此之外，为了配合这一简约的理念，设计师还专门设计了字体和插画，从而赋予包装以独特之美。

Square A B C

Production date: 2010
Designer: Dima Bilan
Client: Supermercado Verdemar
Nationality: Russian

Russian square, everyone should not be unfamiliar. For the design of these paper cups, the designer's inspiration comes from it. This is also what the designer does before a font deformation, and was applied in the cups. When people use it to drink, they can think of Russian square. To everybody's busy life, the cups add a relaxed atmosphere.

"ＡＢＣ方块"

完成时间：2010年
设计师：迪马·比兰
客户：海绿色超级市场
国家：俄罗斯

俄罗斯方块，想必每个人都应该不会陌生。对于这些纸杯的设计，设计师迪马·比兰的灵感正源自于此。同时，这也是一个字体变形前的一个状态。设计师巧妙地将这一方案应用到纸杯之上，促使使用者在喝饮品的同时，对俄罗斯方块进行研究，这一巧妙的设计将为日常忙碌的生活带来一些惬意与放松。

Xmas Coffee Cup

Production date: 2011
Designer: Luciano Ferreira
Client: Dima Bilan
Nationality: Russian

The cup is designed for the design agency's Christmas Party. The designers put the Christmas greetings and the representatives of the simple Christmas graphics on the coffee cups. Everybody in the party would hold the cups and feel the thick Christmas atmosphere.

圣诞咖啡杯

完成时间：2011年
设计师：卢西亚诺·费雷拉
客户：迪马·比兰
国家：俄罗斯

这是设计师专为其设计公司的圣诞派对而设计的纸杯方案。设计师巧妙地将圣诞节的祝福语和极具代表性的圣诞节图案设置在咖啡杯之上。每一位前来参加派对的人士在手握咖啡杯的同时，将能够感受到其中所传递的浓厚的圣诞节气息。

Cross-stitch

Design agency: BD
Production date: 2011
Designer: Damon Scott
Client: cross-stitch
Nationality: British

The cup design mainly comes from a now very popular cross-stitch reflecting a kind of attitude towards life. A lot of inspiration of the designer comes from life. Striking yellow matches with black cross-stitch words, concise and easy, outstanding art from life.

"十字绣"

设计机构：BD设计工作室
完成时间：2011年
设计师：达蒙·斯科特
客户：十字绣
国家：英国

这一纸杯设计的主要灵感源自设计师对当下最受欢迎的"十字绣"的参考。设计师希望通过设计彰显出其一种生活态度，而其设计灵感也大多源自生活。在此，他大胆地选用了亮黄色色调，并搭配以黑色的十字绣风格文字，充分地体现了生活中精炼、简单、出色的艺术。

T Ice Cream

Production date: 2009
Designer: Yanghee Kang
Client: T Ice Cream
Nationality: Korean

T Ice Cream is tea-flavoured ice cream. The word "Tea" and the letter "T" are pronounced the same, so the designer used T for the product name because it arouses more curiosity in customers and leaves a stronger impression. The serif typeface and patterns convey comfortable and elegant feelings. Different colours represent each kind of tea.

T冰淇淋

完成时间：2009年
设计师：梁熙·康
客户：T冰淇淋
国家：韩国

T冰淇淋是一个茶香风味冰淇淋品牌。在英语中，单词"Tea"（茶）和字母"T"的发音相同，因此，设计师选用"T"作为该产品的名称，旨在唤起消费者的好奇心，从而对其留下深刻的印象。衬线字体和图案完美地传达出舒适、优雅之感。不同的色彩代表不同种类的茶味冰淇淋。

Amy's Bread

Production date: 2008
Designer: Yanghee Kang
Client: Personal work
Nationality: Korean

Amy's Bread is one of the most famous bakeries in New York. The logo conveys organic and handmade bread. The round texture represents dough and the Amy's mischievous character reminds customers of their childhood. The overall design gives customers good feelings, such as warm hearts and emotions of closeness.

艾米的面包店

完成时间：2008年
设计师：梁熙·康
客户：个人作品
国家：韩国

艾米的面包店是纽约最著名的面包店之一。独特的标识设计委婉地传递出店内面包的有机选材和手工制作的特点。圆形纹理象征着面团，而"艾米的面包店"的淘气角色则令顾客自然联想起他们的童年。整个设计带给人以美好的感受，所传递的温馨与亲切之感将顾客仅仅地围绕。

Chivalroast Coffee

Production date: 2010
Designer: Charis Ceniseroz
Client: Personal work
Nationality: American

Like a hero in a fairytale, Chivalroast is strong enough to save any day in distress. With bold flavours and romantically illustrated landscapes, the brand and its designs were created to mirror famous legends. As the names suggest, the brand positioned their coffee as the hero, brewed to save your day.

"烘焙骑士"咖啡厅

完成时间：2010年
设计师：查理斯·森赛罗兹
客户：个人作品
国家：美国

犹如童话故事中的一位英雄，"烘焙骑士"咖啡厅每天为疲于奔命的顾客们带来了无限惬意。凭借大胆、独特的口味以及极富浪漫主义格调的插画背景，这一品牌及其设计再次成为一种传奇。正如其名字所指，这一咖啡品牌将成为现实世界中一位伟大英雄，将带走人们每天的奔波疲惫，换来一身的惬意与闲适。

Circ Icecream

Production date: 2009
Designer: Lauren Proctor
Client: Blue Marlin Spark Award Application
Nationality: Australian

Using a vintage circus theme, Circ Icecream gives a sense of childhood nostalgia while representing "good old-fashioned fun". Characters from the circus represent the different flavours. This was the winning submission for the 2009 Blue Marlin Spark Awards.

马戏团冰淇淋

完成时间：2009年
设计师：劳伦·普罗克特
客户：布鲁·马林星火奖应用项目
国家：澳大利亚

设计师劳伦·普罗克特巧妙地选用了一个复古的马戏团主题，从而为这一系列冰淇淋产品增添了些许童年的怀旧之感，并充分展现了"美好的古典意趣"。马戏团中的人物则代表着不同的冰淇淋口味。这一项目是2009年布鲁·马林星火奖的提案。

Yum Dilly

Design agency: Ideas that Kick
Production date: 2009
Designer: Kirsten Swank, Devon Adrian, Carisa Flaherty
Client: Yum Dilly
Nationality: American

From the brand name and identity to the playful illustrations that transform each package into a memorable, cherished keepsake that outlasts the candy inside, WalMart adored Yum Dilly goodies from the start. The Christmas creations were so popular that Valentine's Day and Easter treats were fast-tracked into production.

百胜迪利食品

设计机构：Ideas that Kick设计工作室
完成时间：2009年
设计师：基尔斯滕·斯旺克，德文·阿德里安，凯利撒·弗莱厄蒂
客户：百胜迪利公司
国家：美国

从品牌名称到识别方案再到妙趣横生的插画，精致的细节巧妙地将每一个包装变成了一个难忘、珍贵的纪念品，其影响力大大超越了内置的糖果。沃尔玛超市从一开始即对百胜迪利食品颇有好感。由Ideas that Kick设计工作室设计的圣诞节特别版投放市场之后备受欢迎，而情人节和复活节的特别版设计也随之紧张地展开。

Saugr Moustache Cup

Production date: 2010
Designer: Jacob D'Rozario
Client: Personal work
Nationality: British

Saugr is a brand of organic cane sugar combined with real edible 24k gold flakes. The brand name Saugr, is a play on the word sugar using gold's periodic symbol Au. Upmarket coffee houses had their paper cups rebranded to generate interest about Saugr, as well as giving away free samples of the product.

肖格尔胡须杯

完成时间：2010年
设计师：雅各布·D'罗扎里奥
客户：个人作品
国家：英国

肖格尔品牌强调有机蔗糖与可食用24K金片的成分混合。这一品牌名称"Saugr"是英文单词"sugar"（蔗糖）与黄金的化学符号"Au"的合成体。一些高级咖啡厅在对纸杯进行重新设计的同时，对肖格尔纸杯产生了浓厚的兴趣，同时将该产品的样品免费发放给顾客。

Blow Your Own Trumpet!

Design agency: mattbutterfield
Production date: 2010
Designer: Matt Butterfield
Client: NOISEfestival
Nationality: British

The project is part of an advertising campaign for NOISEfestival.
com, a charity which helps young creatives showcase their work.
The campaign idea was "Blow your own trumpet!" meaning "show
yourself off." When drinking from the cup it appears you are blowing
a mini trumpet, making the advert fun, interesting and amusingly
engaging.

"自己吹小号！"

设计机构：马特·巴特菲尔德设计工作室
完成时间：2010年
设计师：马特·巴特菲尔德
客户：NOISEfestival网站
国家：英国

该项目是NOISEfestival网站的某广告宣传活动中的一个组成部分。
NOISEfestival网站是一个慈善机构，旨在为年轻的创意设计师提供一个展
示他们作品的平台。该活动的理念以"自己吹小号！"为主题，也可理解
为"展现自己"。当使用者在使用这一纸杯品尝饮品时，从外观上来看，
犹如自己在吹一个小号，从而使得整个广告充满意趣和诙谐之感，极具吸
引力。

Chumbi Exotic Ice Cream

Design agency: m.e design
Production date: 2010
Designer: Meg Eaton
Client: Personal work
Nationality: American

Chumbi is a new take on ice cream. Its exotic offerings resulted in a minimilistic design that is countered by the whimsical balancing act of its bold and colourful ingredients and quirky patterned elephant. The simplicity of the packaging leaves the impact on the bold flavour.

春比奇异冰淇淋

设计机构：m.e设计工作室
完成时间：2010年
设计师：梅格·伊顿
客户：个人作品
国家：美国

春比奇异冰淇淋是一个全新的品牌。考虑到这一商品的独特性，设计师巧妙地运用一个简约的设计手法，从而与大胆的色彩元素和离奇的大象图案之间的协调技巧形成强烈的视觉冲击。包装的简约化风格令人将注意力转移到独特的冰淇淋风味之中。

Foxy Foods

Design agency: Channelzero
Production date: 2011
Designer: Imelda Dickinson
Client: Foxy Foods
Nationality: Australian

Pash products use imagery in photographic style from the 1940's & 1950's. The imagery places a quite provocative name in direct juxtaposition with a time of (perceived) innocence and old-fashioned fun. This will ensure that the product appeals to a large target market, is suitable for kids and has a great appeal to start conversations. The colours chosen for the Pash packaging are deliberately rare in FMCG and again, take their inspiration from the 1940's.

小狐狸食品

设计机构：Channelzero设计工作室
完成时间：2011年
设计师：伊梅尔达·迪金森
客户：小狐狸食品
国家：澳大利亚

帕什产品习惯运用20世纪40年代和50年代的摄影图片作为包装的图案。这一图案上还附有极具煽动性的名称，从而与单纯、旧式的有趣时光相得益彰。这一巧妙的设计手法将确保商品面向一个广大的消费者市场，同时也适合于儿童，并为大家提供了极具吸引力的交流话题。帕什产品包装所选用的色彩在快速消费品中很是罕见，其设计风格源自20世纪40年代的特色。

Gelati Sky

Design agency: Truly Deeply
Production date: 2009
Designer: Lachlan McDougall, Cassandra G
Client: Gelati Sky
Nationality: Australian

Gelati Sky is a boutique, premium gelati range. The designers created visuals that were strikingly unique, represented his story and sparked conversation. The concept combined imagery of Italy with objects that represented the flavours creating a unique, organic and scrumptious shape for each flavour. The packaging looks like memories and dreams of Italy.

"冰糕的天空"

设计机构：Truly Deeply设计工作室
完成时间：2009年
设计师：拉克兰·麦克杜格尔，卡桑德拉·G
客户："冰糕的天空"精品店
国家：澳大利亚

"冰糕的天空"精品店以出售优质意大利胶凝冰糕食品为特色。由Truly Deeply设计工作室的设计师所打造的视觉识别方案极其醒目和独特，充分诠释了该店背后的故事，从而引发消费者之间针对这一品牌的交流。设计理念巧妙地将意大利特色形象和代表不同口味的元素完美结合，从而为每种口味塑造了一个醒目独特、结构完整、美妙迷人的形态。整个包装风格呈现出对意大利的怀念与向往之情。

Dress Up Cup & Plate Set

Design agency: rotem peleg graphic design
Production date: 2011
Designer: Rotem Peleg
Client: Personal work
Nationality: Israeli

The disposable cup & plate set turns into a dress up game. The dolls are being cut out of the cups and can be dressed with various clothing and accessories. In addition to the joy of playing, the players can learn something about re-use.

换装杯盘组合

设计机构：罗特姆·皮莱格平面设计工作室
完成时间：2011年
设计师：罗特姆·皮莱格
客户：个人作品
国家：以色列

在这一项目中，一次性纸杯和餐盘组合变成了一个装饰游戏。纸杯上的玩偶被剪下以后，还可以在各种服饰和配饰之中作为装饰之用。这种匠心独运的设计手法，不仅极具创作乐趣，同时也帮助用户体会"再利用"的含义。

The Nations

"联合国"

Production date: 2011
Designer: Alex Litovka
Client: Personal wprk
Nationality: Belarussian

完成时间：2011年
设计师：亚历克斯·里图维克瓦
客户：个人作品
国家：白俄罗斯

The main idea was to create original cups which will symbolise different nations. The designer decided to choose these nations: English (Policeman), Dutch (Girl in national costume) and Mexican (Man in poncho). This project was a part of university homework. The task was to create original package consisting of three things.

该项目的主要理念是创造出若干匠心独运的纸杯，并使之代表不同的国家。设计师亚历克斯·里图维克瓦最终选定了如下几个国家：英国（警察）、荷兰（穿民族服装的少女）、墨西哥（穿斗篷的人）。该项目是一个大学作业的一个部分。其主要任务是设计一款独一无二的包装，要求其中具备三种元素。

Café Opus

Design agency: Sukker Design
Production date: 2011
Designer: Hanna Heiness, Ellen Brusgard, Alexander Qual
Client: Umoe Restaurant Group AS
Nationality: Norwegian

It is a complete redesign of the largest shopping centre café/restaurant chain in Norway. The designers developed the full brand experience from conceptual development, identity, packaging to guide lines for interior design.

欧普思咖啡馆

设计机构：苏克尔设计工作室
完成时间：2011年
设计师：汉娜·海内斯，艾伦·布鲁斯戈尔德，亚历山大·库尔
客户端：Umoe餐厅集团
国家：挪威

该项目是苏克尔设计工作室专为挪威最大的购物中心咖啡馆/餐厅而设计的品牌重塑方案。设计师们开发的全面品牌设计方案包括：品牌概念的开发、识别设计、包装设计以及室内设计的指导方针的确立等。

Frost

Production date: 2009
Designer: Clara Tan
Client: Personal wprk
Nationality: Singaporean

This is a branding project for Frost, a make-believe ice cream parlour that has a fun and quirky personality. Its startup is based on the belief that eating ice cream should be an enjoyable experience, rather than just eating them as mere desserts. From buying ice cream to eating them in a fun, comfortable environment, enjoying ice cream should be a lifestyle.

弗罗斯特冰淇淋

完成时间：2009年
设计师：克莱尔·谭
客户：个人作品
国家：新加坡

"弗罗斯特"是一家有趣而奇特的冰淇淋店，而这一项目是设计师专为其设计的品牌方案。设计从一个理念出发，即品尝冰淇淋必定是一次快乐的享受，而不仅仅是一个甜点所带来的味觉刺激。从购买冰淇淋到在一个有趣、舒适的环境中品尝冰淇淋乃至全心身地享受这一过程，其实应该是一种生活方式。

汉堡国王全球包装方案

设计机构：克里斯宾·波特+伯格斯基设计工作室
完成时间：2010年
设计师：戴维·伊格莱西亚斯
客户：汉堡国王
国家：美国

2010年，汉堡国王公司发布了其最新的改良
版全球包装方案。这一快餐业巨头力图在全球
的多样化市场上扩张其影响力，尽量减少与其
他同类商品相似的可能性，并有效地突出其产
品的特色。最终的设计方案是设计师运用了一
个大胆的说明性设计手法，充分展现了该公司
菜单中最受欢迎的项目。

Burger King Global

Design agency: Crispin Porter + Bogusky
Production date: 2010
Designer: David Iglesias
Client: Burger King
Nationality: American

In 2010, Burger King launched its newly reinvented global packaging. The fast-food king sought to universally expand its appeal across various international markets by minimising the use of copy and adding more emphasis to their products. The result was a bold, illustrative approach at showcasing some of their more popular menu items.

Eat Pastry Vegan Cookie Dough

Design agency: Chelsea Koornick Designs
Production date: 2009
Designer: Chelsea Koornick
Client: Eat Pastry
Nationality: American

Eat Pastry is a vegan cookie dough company that wanted to stand out from the rest of the ready-made cookie doughs on the grocery store shelves. The designer wanted to create a design with a retro yet modern feel, heavily focused on typography; something that hasn't quite been done before in this particular product niche. This project is only a concept piece.

乐享糕点 —— 素食饼干面团

设计机构：切尔西·库尔尼克设计工作室
完成时间：2009年
设计师：切尔西·库尔尼克
客户：乐享糕点
国家：美国

乐享糕点是一个素食饼干面团加工公司，希望通过醒目的品牌设计能够在杂货店货架上独领风骚，从而拉开与其他同类素食饼干面团产品的距离。因此，设计师力创建一个集复古和时尚风格于一体的设计方案，重点突出字体的设计；除此之外，在这一特殊产品定位之前还有很多程序需要进行，而该项目仅仅是设计概念的一个部分。

Eat Right

"巧食"

Production date: 2011
Designer: Yotam Bezalel
Client: Eat Pastry
Nationality: Israeli

完成时间：2011年
设计师：约塔姆·比扎莱尔
客户：乐享糕点
国家：以色列

Eat Right's exclusivity which sets them apart from the other chains is the claim that their products are 100% healthy. The designers chose to create a "coacher" character that represents a combination of professionalism, transparency along with a bit of humour. The coacher accompanies the costumer on every printed product and gives him tips for a healthier and more balanced way of life. In order to emphasise the value of health, they designed a special icon language (an icon for each nutritional value) that is used as an info-graphic on the printed products of the chain.

"巧食"在众多连锁店中独占鳌头的一个主要原因是他们所倡导的100％健康食品理念。设计师约塔姆·比扎莱尔等人精心塑造了一个"教练"的形象，使之成为各种专业精神的合成体，透明并伴有一丝幽默感。这一"教练"出现在每一个印刷产品之中，伴随着顾客，并为他们提供一个健康而更为平衡的生活方式。为了突出健康的价值，设计师还精心设计了一个特殊的图标语言（一个图标代表一种营养价值），并将其用作这一连锁店印刷产品上的一个信息图形。

Fresssup

Design agency: kissmiklos
Production date: 2010
Designer: Miklós Kiss
Client: Fresssup
Nationality: Hungarian

This is a cup design fora fresh wringed drinks company. It also includes name, slogan, package design, uniform and ads graphic concept design.

Fresssup公司

设计机构：kissmiklos设计工作室
完成时间：2010年
设计师：米克洛什·吉斯
客户：Fresssup公司
国家：匈牙利

这一系列纸杯由设计师米克洛什·吉斯专为一个经营鲜榨饮品的公司而设计。除此之外，设计师还为该公司提供了品牌名称、包装、制服以及广告平面概念设计方案。

Two Seasons Coffee

Design agency: Barker Gray
Production date: 2011
Designer: Pan Yamboonruang, Claire Casey
Client: Two Seasons Coffee
Nationality: Australian

The aim of the design is to create customer & consumer interest, awareness and ultimately trial of the newly developed Two Seasons coffee brand. The designers had the full support of the client team to break all the traditional codes of roast & ground coffee language (black, gold, heritage credentials, etc.) that persists in the many upmarket and boutique cafés in metropolitan Australia. They wanted to drive in the opposite direction of gourmet coffees overriding archetype "The Lover". Two Seasons breaks all convention and is purposefully everything we don't know "origin"-based coffees to be... joyful, colourful and fun... the promise of a great experience to come, which for those who love their coffee it delivers.

"两季咖啡"

设计机构：巴克尔·格雷设计工作室
完成时间：2011年
设计师：潘·亚姆波鲁朗，克莱尔·凯西
客户：两季咖啡厅
国家：澳大利亚

该项目的设计目的是为一个新建立的"两季咖啡"品牌创建一个令客户与消费者满意、理解并认可的品牌方案。在客户团队的大力支持下，巴克尔·格雷设计工作室勇于打破澳大利亚大都市中诸多高档精品咖啡馆中传统的烘焙&研磨咖啡语言代码（黑色、金色、传统认证等）。设计师们希望改变传统的咖啡厅品牌设计路线，对"最爱"的原型进行重新定义。"两季咖啡"的设计突破了所有传统的束缚，向人们证明咖啡本身应该是快乐、多彩和乐趣的，而这恰恰是人们之前不曾发现的。两季咖啡厅将带给那些真正了解咖啡的人士一个伟大的感官体验。

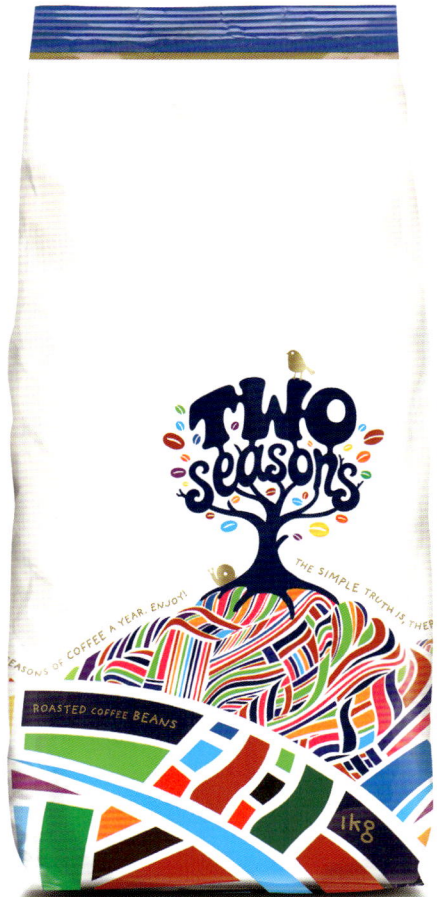

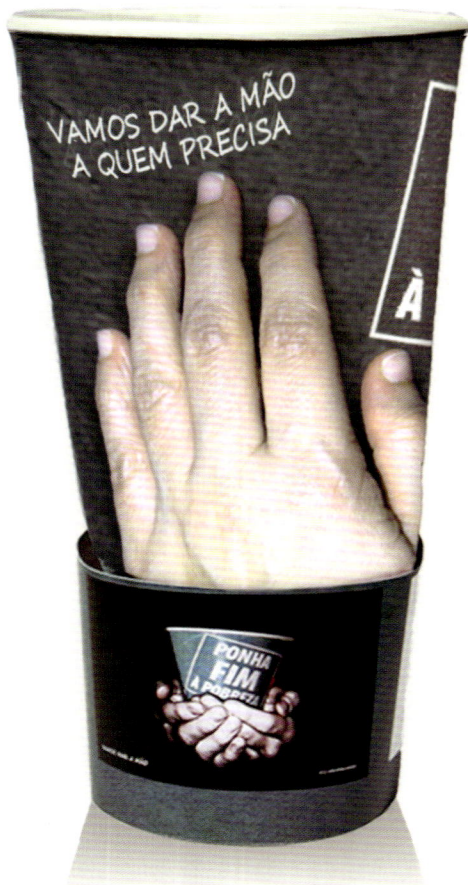

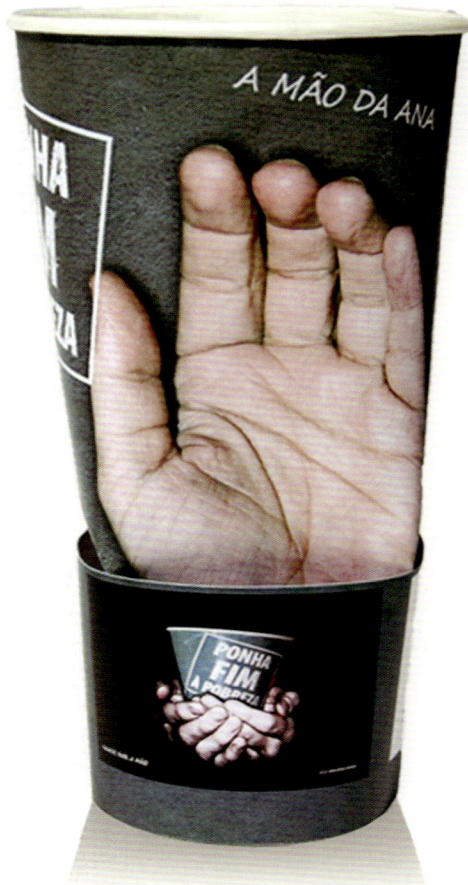

Give a Hand

Design agency: Ana Silvia Santos Design
Production date: 2010
Designer: Ana Silvia Santos
Client: 2010 European Year for Combating Poverty and Social Exclusion
Nationality: Portuguese

The hand as a profound effect, not only communicative but the join of hands to pray asks support and gives something. The join of hands to ask, which forms the cup design, helps to mark the year 2010 as European Year of Combating Poverty and Social Exclusion. The cup has no support base, as people who need support and it's safe by the information brochure.

"提供帮助"

设计机构：安娜·西尔维娅·桑托斯设计工作室
完成时间：2010年
设计师：安娜·西尔维娅·桑托斯
客户：2010年欧洲贫穷与社会排斥斗争年
国家：葡萄牙

具有深刻影响力的手，不仅是沟通、交际的重要手段，同时也是祈祷、支持、给予的一种象征。双手交叠的形式构成了这一纸杯的设计方案，有力地突出了2010年欧洲贫穷与社会排斥斗争年的主题。这一纸杯并没有支撑的基底，象征着需要被支持的团体以及资料小册子的可靠性。

232

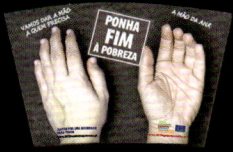

Hummm

Design agency: joaoricardomachado
Production date: 2011
Designer: João Ricardo Machado
Client: Hummm Café
Nationality: Spainish

The design is a package family for the Hummm Ice Cream Store. "There's this beautiful land where everything is sweet and the rain drops are made of Ice Cream, Milkshake, Coffee and Chocolate. Hummm". It is very bright, fresh and colourful to make life happier.

Hummm咖啡店

设计机构：若昂·里卡多·马查多设计工作室
完成时间：2011年
设计师：若昂·里卡多·马查多
客户：Hummm咖啡店
国家：西班牙

这是设计师若昂·里卡多·马查多专为Hummm冰淇淋店而设计的家庭装包装方案。"有这样一个美妙的国度，那里的一切都是甜美的，即便是雨点中也混合着冰淇淋、奶片、咖啡巧克力，这就是Hummm咖啡店。"醒目、清新、缤纷的配色方案使生活也变得格外幸福。

Café De Luxe

奢华咖啡厅

Production date: 2007
Designer: Ka Yin Karen Lam
Client: Café De Luxe
Nationality: Australian

完成时间：2007年
设计师：卡·林·凯伦·林姆
客户：奢华咖啡厅
国家：澳大利亚

The aim of this project is to create a distinctive and stylish brand identity and stationery applications for this new modern café in Canberra, Australia. The café's owners want to give their customers cheerful experiences when they come to their café. Therefore, the idea of using colourful stripes is to meet this particular purpose and also interpret there are different kinds of coffees and food in this café.

该项目的目的是为澳大利亚堪培拉新落成的一个现代咖啡厅创建一个独一无二、风格独具的品牌识别与文具应用方案。该咖啡厅的主人希望能够为进入空间的顾客营造出一个令人雀跃的空间体验。因此，设计师选择了彩色条带的创意理念以满足这一客户的要求，并委婉地体现出店内所出售咖啡和食品种类的繁多。

Soyato

Design agency: Acacia Design Consultants
Production date: 2010
Designer: Kelvin Ng
Client: Soyato
Nationality: Singaporean

By bringing together expert knowledge in food science, a love affair with desserts, and a large dose of fun, Soyato's mission is to create a new category and brand of frozen dessert based on the benefits of soy – that will enable the "Soyato Clan" to lead the healthy, fun-filled lifestyle they desire!

Soyato冰淇淋

设计机构：金合欢设计顾问有限公司
完成时间：2010年
设计师：凯尔文·尼格
客户：Soyato冰淇淋
国家：新加坡

Soyato冰淇淋店将其针对食品科学的专业知识、对甜点的热爱之情以及诙谐风趣的个性完美地融为一体。它创建了一个全新的范畴，以黄豆为原料的这一冰冻甜点品牌将引领一个健康、有趣的生活方式。

So Espresso Cup

Design agency: Mousegraphics
Production date: 2011
Designer: Vassiliki Argyropoulou
Client: Draculi Coffee
Nationality: Greek

So Urgent.
So tasty.
So strong.
So unique.
I am so saved...
If the man is his coffee, then... So be it...
At your disposal.

"如此"咖啡杯

设计机构：Mousegraphics设计工作室
完成时间：2011年
设计师：瓦西莉基·阿尔戈罗普罗
客户：德拉库里咖啡
国家：希腊

如此迫切。
如此美味。
如此强烈。
如此独特。
此刻，再多的华丽辞藻也变得多余……
如果将人视为一杯咖啡，那么……那就这样吧……
任你自由支配。

Magic Surprise Ice Cream

Production date: 2010
Designer: Jonathan Capecch
Client: Cardonald
Nationality: Scottish/British

"惊奇魔术"

完成时间：2010年
设计师：乔纳森·凯普科
客户：Cardonald冰淇淋
国家：苏格兰/英国

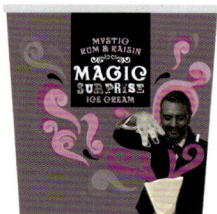

Jonathan was asked to create a new brand of ice cream which would challenge the market leaders but be a cheaper alternative to the likes of Haagen-Dazs or Ben & Jerry's. His challenge was to give the brand maximum shelf impact when alongside these better known brands but to offer something new and different as well. Magic Surprise was the concept he came up with, gaining inspiration from magic shows and posters from the 1920's and 30's.

设计师乔纳森·凯普科受客户的委托设计一款全新的冰淇淋品牌，从而使之能够媲美市场上的同类主打产品，但是价位略低于哈根达斯或本&杰里品牌。设计的挑战在于如何扩大该产品在货架上的影响力，打败其他众多知名品牌，并彰显出其与众不同的个性魅力。"惊奇魔术"是设计师最终制定的设计理念，这一理念取材自20世纪20、30年代的魔术表演和海报。

Misr Café

Production date: 2010
Designer: Sina Chakoh, Sebastian Kösters, Kathrin Böhlke, Selin Estroti
Client: Misr Café
Nationality: German

This project is the redesign for the Egyptian coffee company Misr Café. The target was to create a design concept that would be successful in Eastern and Western countries. The motto "flavour your friendship" creates a personal relationship between the client and the product. The logo is remarkable, clear and modern including a whiff of orient.

Misr咖啡馆

完成时间：2010年
设计师：辛那·查克哈，塞巴斯蒂安·库斯特斯，凯萨琳·博科，塞林·艾斯特洛里
客户：Misr咖啡馆
国家：德国

该项目是设计师专为埃及咖啡公司旗下的Misr咖啡馆而设计的品牌重塑方案。其目标是创建一个适用于东西方国家的设计方案。店内的宣传口号"为你的友谊增添点乐趣"，有效地在顾客和产品之间建立起一个私人关系。标识的设计十分醒目、清晰、时尚，夹杂着些微的东方韵味。

Mug Packaging Travis Pastrana Limited Edition 特拉维斯·帕斯特拉纳限量版纸杯包装

Design agency: SpiralMotion
Production date: 2010
Designer: Jhonny Bello
Client: University Jorge Tadeo Lozano
Nationality: Colombian

设计机构：螺旋运动设计工作室
完成时间：2010年
设计师：乔尼·贝洛
客户：豪尔赫·塔德奥·洛萨诺大学
国家：哥伦比亚

This project was submitted for an open call from the University Jorge Tadeo Lozano with the purpose to promote the motocross sport, and specifically the motorcyclist "Travis Pastrana". The support of this promotion was the packaging protector used for cups of hot drinks, and thus the designers decided to use coffee since it is the national drink of Colombia. The concept of this project was developed based on the idea of the "dirt" generated by the motocross sport, combining this idea with the aesthetics of the stains generated by the coffee. Subsequently, a digital collage was performed using a contrasting white with dark brown, with figures in high contrast and exploring the coffee stain as an ornament. The package, in addition to promoting the sport of motocross and the sportsman, was used to promote coffee as the national drink of Colombia.

该项目是豪尔赫·塔德奥·洛萨诺大学公开选拔设计大赛的一个提交项目。这一大赛的举办意在有效地宣传摩托车越野赛以及启动引擎车手"特拉维斯·帕斯特拉纳"。作为这一大赛的一个辅助项目，设计师力图设计一款防烫杯套，并赋予其哥伦比亚特色。该项目的设计理念是以摩托车越野赛产生的"尘土"为基础，巧妙地将这一理念与咖啡污渍的美学理念完美结合。随后，设计师使用对比鲜明的白色与深棕色色调，搭配高对比度的数字，从而形成一个数字拼贴图案，并将咖啡污渍开发成一种装饰元素。这一精致独特的包装不仅对摩托车越野赛和运动员进行了有效的宣传，同时也促进这一咖啡成为哥伦比亚的国饮。

OrangeCup Natural Frozen Yogurt

Design agency: Range
Production date: 2011
Designer: John Swieter, Garrett Owen
Client: OrangeCup
Nationality: American

The OrangeCup launch kit was created to promote the brand personality upon each store opening while reinforcing the brand through the use of the colour "orange". The launch kit messaging focused on the simplicity of the product ingredients. Pure. Natural. Perfect. The natural solution for the launch kit uniforms and packaging was to label them for exactly what they are.

香橙杯天然冷冻酸奶

设计机构：范围设计工作室
完成时间：2011年
设计师：约翰·斯威特，欧文·加勒特
客户：香橙杯天然冷冻酸奶
国家：美国

香橙杯天然冷冻酸奶品套餐的精心设计，在每个店面中体现了品牌的特色，通过"橙色"的运用借以强化这一品牌的主题。这一系列产品的包装强调产品成分的纯天然，即纯粹、天然和完美。除此之外，该系列制服与包装的纯天然设计方案也有效地突出了产品的本质。

Polka Gelato

Design agency: VONSUNG
Production date: 2011
Designer: Joseph Sung
Client: Michiko Ito
Nationality: Korean

波尔卡意式冰淇淋店

设计机构：VONSUNG设计工作室
完成时间：2011年
设计师：约瑟夫·孙
客户：伊藤道子
国家：韩国

VONSUNG recently completed the total identity design for Polka Gelato. Based in a conservation area, Fitzroy Square, Polka Gelato opens its doors to showcase their artisanal way of creating ice cream. Despite all the talk of a double-dip recession in the UK, the client's wish was to offer something enlightening, from old to young, a sense of affordable luxury amid these difficult times.

VONSUNG设计工作室最近完成了波尔卡意式冰淇淋店的全部识别方案的设计。波尔卡意式冰淇淋店坐落在菲茨罗伊广场的一个保护区内，旨在面向广大顾客展现其精湛的冰淇淋工艺加工技巧。在英国出现经济衰退的大环境下，客户希望他们的品牌设计方案能够带给顾客一定的启发，无论是年长者还是青年人，均能够在这一经济困难时期享受到一种负担得起的奢华体验。

Bom Dia! Vida e Caffè

Design agency: Simon Kuhn Design
Production date: 2011
Designer: Simon Kuhn
Client: Vida e Caffè
Nationality: South African

This is a submission for the ten year birthday, limited edition coffee cup sleeve for Vida e Caffè, a chain of Portuguese/South African coffee bars. The design incorporates the brand's slogan, "Bom Dia!" which means "Good Day" in Portuguese, emblazoned across the knuckles of a fighter, who "pities the fool". This was used to emphasise the powerful punch of the coffee while incorporating a "retro" boxing poster aesthetic.

"好日子！" —— 维达咖啡厅

设计机构：西蒙·库恩设计工作室
完成时间：2011年
设计师：西蒙·库恩
客户：维达咖啡厅
国家：南非

该项目是西蒙·库恩设计工作室专为维达咖啡厅的十周年纪念日而设计的限量版咖啡杯杯套。维达咖啡厅是葡萄牙/南非的一个连锁咖啡吧。该设计巧妙地将这一品牌的宣传口号 "Bom Dia!"（在葡萄牙语中寓意 "好日子"）添加到绘有拳击手（"同情傻瓜"）双拳的图案之中。这一精心的设计旨在突出咖啡机的强大冲压力，并结合一个 "复古的" 拳击海报美学理念。

Pour Moi Paper Cups

珀尔·莫伊纸杯

Design agency: SMR Creative Agency
Production date: 2010
Designer: Nicola Johnston
Client: Aimia Foods Limited
Nationality: British

设计机构：SMR创意公司
完成时间：2010年
设计师：尼古拉·约翰斯顿
客户：艾米亚食品有限公司
国家：英国

SMR Creative had to create a range of limited edition paper cups to highlight the premium coffee concept vending brand "Pour Moi". Each cup presents an offbeat coffee fact in a contemporary style, creating a talking point for consumers.

这一项目是SMR创意公司为珀尔·莫伊品牌创建的一系列限量版咖啡杯，旨在更好地突出这一品牌咖啡的优质。其中，每款纸杯均以一个时尚的风格进行设计，以展现出每种咖啡口味的独一无二，并为顾客提供一个谈论的焦点。

SOBO Ice Cream

Design agency: Alistair Stephens Design
Production date: 2009
Designer: Alistair Stephens
Client: SOBO Chocolate
Nationality: British

A modest cookie maker in London wanted a modern twist to their new ice cream products. The style of type changes on each label to give the flavours a sense of individuality and makes a distinctive impression in-store.

SOBO冰淇淋

设计机构：阿利斯泰尔·斯蒂芬斯设计工作室
完成时间：2009年
设计师：阿利斯泰尔·斯蒂芬斯
客户：SOBO巧克力
国家：英国

SOBO，这位来自英国的谦逊的饼干制造商委托阿利斯泰尔·斯蒂芬斯设计工作室为其新推出的冰淇淋产品打造一个时尚的包装方案。每个标签上的字体风格均富于变化，彰显出与众不同的个性，同时也为店铺增添了独一无二的视觉效果。

Tusso Cup

Design agency: Mousegraphics
Production date: 2009
Designer: Vassiliki Argyropoulou
Client: Draculi Coffee
Nationality: Greek

Unique simplicity, premium aesthetics and product quality are the main principles of Tusso products and the designers want this to be obvious to the customers. The black & white photos of unique people/figures, each chosen for a particular product filled with solid colour, matches the personality of each product. Simple forms, clear typography and the use of black for background colour, simply stability and confidence to the products' quality.

塔索纸杯

设计机构：Mousegraphics设计工作室
完成时间：2009年
设计师：瓦西莉基·阿尔戈罗普罗
客户：德拉库里咖啡
国家：希腊

简单独特、优质高端的美学理念与产品质量是塔索品牌产品的主要经营原则。设计师力图通过设计将这一原则完美呈现给顾客。每张以特定人物或数字为背景的黑白照片均代表一个特殊的产品，象征着每种产品的独特性。简单的模式、清晰的字体、黑色的背景完美结合，稳重而真诚地传达出产品的优质。

Voglia di Gelato

Design agency: Basile Advertising
Production date: 2010
Designer: Andrea Basile
Client: Gelati Aloha srl
Nationality: Italian

Voglia di Gelato is a line of Italian ice cream. The top and body of each package are decorated with simple, one colour graphics, different for each flavour of the ice cream. It is subdivided among flavours of fruit sorbet and ice cream. The pleasure to enjoy the real homemade ice cream at home.

"渴望冰淇淋"

设计机构：巴西莱广告创意公司
完成时间：2010年
设计师：安德烈·巴西莱
客户：阿罗哈冰淇淋公司
国家：意大利

"渴望冰淇淋"是一个意大利冰淇淋系列产品品牌。每个包装的杯盖和杯体均饰有简约的单色图案，不同口味的冰淇淋图案也不尽相同。同时，精巧细致的设计还巧妙地将水果冰沙和冰淇淋风味进行细分。愉悦的感官体验令人自然地联想起家中自制冰淇淋的味道。

Be Different!

Production date: 2011
Designer: Madeline Lim
Client: Udders
Nationality: Singaporean

This is a rebranding project for Udders, a local ice cream which liqueur ice cream as their specialty. Its brand personality is different, fun, quirky, weird and unique; hence a quirky-looking yellow cow entirely made up of shapes is created as a primary graphic image for the brand. It looks edgey and unique, to give Udders a brand new identity and look. The secondary graphic image is simply a pyramid inspired by ice cream cones. With those two images in mind, both the designs for the pint and cup are formed by repeating the cow and the pyramid all over. The cows are positioned in different positions and angles, to make the whole product look more lively, fun and quirky.

"就是不同！"

完成时间：2011年
设计师：马德琳·林姆
客户：Udders冰淇淋店
国家：新加坡

该项目是设计师专为Udders冰淇淋店而打造的品牌重塑方案，作为一个地方冰淇淋店，Udders以经营利口酒冰淇淋为特色。其品牌个性独特、风趣、狡黠、奇异，因此，设计师马德琳·林姆最终塑造了一个造型不一的黄牛形象作为该品牌的主题图案。这一形象前卫而独特，使Udders品牌的识别和形象焕然一新。此外，另一个附属图案是一个简约的金字塔，取材自甜筒的外形。这两个主要图案遍及整个印刷材料和纸杯的设计之中。黄牛位置和角度的变换为整个商品增添了生动、诙谐、个性化的气息。

Yummy Ice Cream

Design agency: joaoricardomachado
Production date: 2011
Designer: João Ricardo Machado
Client: Yummy Ice Cream
Nationality: Portuguese

This is a project for an ice cream identity and package family for kids. The faces expressions interacts with the kids by showing how fun and delicious is to eat ice cream and turn the package into a collectible and reusable object for kids' fun. The colours were chosen based on the real colour of each ice cream flavour making it easy to recognise. These vibrant colours give more energy to the faces and look very fun to the kids. When you have the happy face turned to you, all you can see is this funny and cute face smiling and calling you to see what is there. Particularly, the designer always feels like laughing when he looks at these happy faces.

美味冰淇淋

设计机构：若昂·里卡多·马查多设计工作室
完成时间：2011年
设计师：若昂·里卡多·马查多
客户：美味冰淇淋
国家：葡萄牙

该项目是若昂·里卡多·马查多设计工作室设计的一个冰淇淋识别设计和儿童家庭装包装方案。纸杯上的笑脸将有效地促进商品与儿童的互动，向他们展现品尝这一美味冰淇淋的喜悦之情，同时也可以成为孩子们一个可收集和再利用的玩具。除此之外，设计师巧妙地运用与冰淇淋原色接近的色彩，从而使各种口味的冰淇淋更加易于辨识。这些充满活力的色调为每张笑脸增添了无限活力，并能够轻松捕捉孩子们的眼球。无论在哪里，一张向你微笑的脸，定会吸引你的靠近。每次设计师自己面对这些笑脸时，都会不自觉地发笑。

Paper Cups for VNU Media

荷兰联合出版集团传媒的纸杯设计

Design agency: kvhw.nl
Production date: 2011
Designer: Julia Kaiser
Client: VNU Media
Nationality: Dutch

设计机构：kvhw.nl设计工作室
完成时间：2011年
设计师：朱莉娅·凯撒
客户：荷兰联合出版集团传媒
国家：荷兰

The design is to activate the six business values of VNU Media: openness, customer focus, teamwork, drive, entrepreneurship and professionalism. The solution? Six catchy mentality statements in bright corporate colours: "Everything is open" for openness; "There is only one boss" for customer focus; "1+1 = 11" for teamwork; "Winning with double I" for drive; "This is your chance" for entrepreneurship and "Till here and further" for professionalism. All six statements were applied to various communication tools, amongst others to the company's paper cups. The result? More focus, more orientation, more pride, more connection. In short an interpretation of the business values that really matters.

该项目的设计目的在于有效地激活荷兰联合出版集团传媒的六个企业价值观，即开放、客户至上、团队合作、动力、创业精神和专业精神。那么，最好的解决方案是什么呢？设计师巧妙运用鲜明的色调将这六个价值观浓缩成相应的语句，醒目而极富感染力。这六个语句分别是："所有的一切都是开放的"代表"开放"；"只有一个老板"代表客户至上；"1+1 = 11"代表团队合作；"以两个I为根本"代表动力；"这是你的机遇"代表创业精神；"立足当下，面向未来"代表专业精神。这六种陈述被应用到所有的传达工具之中，同时也包括公司的纸杯。反响如何？得到了更多的关注、更多的方向、更多的骄傲、更多的连接。总之，将企业价值观进行简洁的诠释具有十分重要的意义。

252

INDEX
索引

Justin Marimon
Ka Yin Karen Lam
Kareem Gouda
Katie Lewis
Katrine Austgulen
Kelsey McNabb
Kevin Leung
Kissmiklos
Klaus Voorman
Kristin Hubbard
Kvhw.nl
Leaf Design Pvt. Ltd.
Luciano Ferreira
Luko Designs®
M.e design
Madeleine Ward, Nicole Powell
Madeline Lim
Maite Cantó
Manara Design Studio
Mats Ottdal Design
Mattbutterfield
Millie Rose Cordingley
MIN Design Studio
Mitre Agency
Mohammed Al-Mousa
Moodley brand identity
Mousegraphics
Mr. Conde Studio
Noahbutcher
Noote&Netoo
Novel
OBAH Design
O-D Studio
Olly Blake
Paprika Design LTDA
Petty Hartanto
Phil Héroux
Phuong Ngoc Le
Piraya Ruangpungtong

Priz Praz Pruz
Range
Ricky Martin
Rotem peleg graphic design
Savannah College of Art and Design
Seth Beukes
Simon Kuhn Design
Sina Chakoh, Sebastian Kösters, Kathrin
Böhlke, Selin Estroti
SMR Creative Agency
Sole Designer
SpiralMotion
Staffordshire University
Sterling Brands
Studio Muti
Studio43
Sukker Designer
Sveta Fedarava
Tanja Doepke
Tara Dayne Walbridge
The Creative Circus
Thing Studio
Tinytwiggette Design
Tony Stavrou
Truly Deeply
Urbe
Vera Valentine
Vincent Van Gogh
Vltor Lopes Leite, Ramon Villain Santos,
Laio de Carvalho, Hellen Aquino Martins
VONSUNG
WeAreAllConnect
Wernery.com
Wink
WPIXY STUDIO
Yanghee Kang
Yotam Bezalel
Zaven

图书在版编目（CIP）数据

纸杯设计 / （波兰）伊维莉娜·柏臣编；贺丽译. --
沈阳：辽宁科学技术出版社，2013.3
ISBN 978-7-5381-7834-0

Ⅰ．①纸… Ⅱ．①伊… ②贺… Ⅲ．①水－生活用
具－设计 Ⅳ．①TS972.23

中国版本图书馆CIP数据核字（2013）第003048号

出版发行：辽宁科学技术出版社
　　　　　（地址：沈阳市和平区十一纬路29号　邮编：110003）
印 刷 者：利丰雅高印刷（深圳）有限公司
经 销 者：各地新华书店
幅面尺寸：170mm×220mm
印　　张：16
字　　数：50千字
印　　数：1～3000
出版时间：2013年 3 月第 1 版
印刷时间：2013年 3 月第 1 次印刷
责任编辑：陈慈良
封面设计：周　洁
版式设计：周　洁　陆浩洋
责任校对：周　文
书　　号：ISBN 978-7-5381-7834-0
定　　价：78.00元

联系电话：024-23284360
邮购热线：024-23284502
E-mail: lnkjc@126.com
http://www.lnkj.com.cn
本书网址：www.lnkj.cn/uri.sh/7834